T0302234

Novel Materials and
Devices for Spintronics

MATERIALS RESEARCH SOCIETY
SYMPOSIUM PROCEEDINGS VOLUME 1183

Novel Materials and Devices for Spintronics

Symposium held April 14–17, 2009, San Francisco, California, U.S.A.

EDITORS:

Stefano Sanvito
Trinity College
Dublin, Ireland

Olle Heinonen
Seagate Technology
Bloomington, Minnesota, U.S.A.

Valentin Alek Dediu
ISMN-CNR
Bologna, Italy

Nick Rizzo
Everspin Technologies
Chandler, Arizona, U.S.A.

Materials Research Society
Warrendale, Pennsylvania

CAMBRIDGE
UNIVERSITY PRESS

University Printing House, Cambridge CB2 8BS, United Kingdom

One Liberty Plaza, 20th Floor, New York, NY 10006, USA

477 Williamstown Road, Port Melbourne, VIC 3207, Australia

314-321, 3rd Floor, Plot 3, Splendor Forum, Jasola District Centre, New Delhi - 110025, India

79 Anson Road, #06-04/06, Singapore 079906

Cambridge University Press is part of the University of Cambridge.

It furthers the University's mission by disseminating knowledge in the pursuit of education, learning and research at the highest international levels of excellence.

www.cambridge.org
Information on this title: www.cambridge.org/9781605111568

Materials Research Society
506 Keystone Drive, Warrendale, PA 15086
http://www.mrs.org

First published 2009
First paperback edition 2012

Single article reprints from this publication are available through University Microfilms Inc., 300 North Zeeb Road, Ann Arbor, MI 48106

CODEN: MRSPDH

A catalogue record for this publication is available from the British Library

ISBN 978-1-605-11156-8 Hardback
ISBN 978-1-107-40821-0 Paperback

This work was supported in part by the Army Research Office under Grant Number W911NF-09-1-0097. The views, opinions, and/or findings contained in this report are those of the author(s) and should not be construed as an official Department of the Army position, policy, or decision, unless so designated by other documentation.

PREFACE

Very few research fields have moved from being a lab curiosity to massive markets as quickly as spintronics. This was celebrated in 2007 with the Nobel Prize awarded to Albert Fert and Peter Grünberg. In the last few years the range of materials used for spin-devices has witnessed a dramatic broadening, now including both organic molecules and multifunctional compounds as well as the standard metals and inorganic semiconductors. Symposium FF, "Novel Materials and Devices for Spintronics," held April 14–17 at the 2009 MRS Spring Meeting in San Francisco, California focused on both the experimental and theoretical aspects of spin phenomena in the solid state with particular emphasis on traditional (inorganic metals and semiconductors) and novel materials like organics and multiferroics.

The symposium organizers would like to thank the staff at the Materials Research Society for their help and support in organizing the symposium. Financial support from the Materials Research Society and the Army Research Office is gratefully acknowledged.

Stefano Sanvito
Olle Heinonen
Valentin Alek Dediu
Nick Rizzo

August 2009

MATERIALS RESEARCH SOCIETY SYMPOSIUM PROCEEDINGS

MATERIALS RESEARCH SOCIETY SYMPOSIUM PROCEEDINGS

Prior Materials Research Society Symposium Proceedings available by contacting Materials Research Society

Dilute Magnetic Semiconductor
and Oxides

Mater. Res. Soc. Symp. Proc. Vol. 1183 © 2009 Materials Research Society   1183-FF01-07

On the defect induced ferromagnetic ordering above room-temperature in undoped and Mn doped ZnO thin films

Mukes Kapilashrami[1], Jun Xu[1,2], Valter Ström[1], K.V. Rao[1], Lyubov Belova[1]

[1]Department of Materials Science, Royal Institute of Technology, Stockholm, Sweden
[2]State Key Laboratory of Materials Modification, Department of Physics, Dalian University of Technology, Dalian, China

ABSTRACT

Evidence for long range ferromagnetic order above room-temperature, RTFM, in pristine ZnO, In_2O_3, TiO_2 nanoparticles and thin films, containing no nominal magnetic elements have been reported recently. This could question the origin of RTFM in doped dilute alloys if for example the ZnO matrix itself develops a defect induced magnetic order with a significant moment per unit cell. In this presentation we report a systematic study of the film thickness dependence of RTFM in pure ZnO deposited by DC Magnetron Sputtering. We observe a maximum in the saturation magnetization, M_S, value of 0.62 emu/g (0,018 μ_B/unit cell), for a ~480 nm film deposited in an oxygen ambience of appropriate pressure. Above a thickness of around 1 μm the films are diamagnetic as expected. We thus see a sequential transition from ferromagnetism to para- and eventual diamagnetism as a function of film thickness in ZnO. We also find that in such a ZnO matrix with a maximum intrinsic defect induced moment, on doping with Mn the maximum enhanced M_S value of 0.78 emu/g is obtained for 1at.% Mn doping. With this approach of appropriate doping in a defect tailored matrix, we routinely obtain RTFM in both undoped and Mn- doped ZnO thin films.

INTRODUCTION

In recent years room temperature ferromagnetism, RTFM, has been reported in otherwise diamagnetic metal oxides (MO) in the form of nanoparticles, nanorods and thin films of thickness comparable to that of the particle size. Saturation magnetization values spreading over a wide range, 10^{-4}-10^{-1} emu/g, have been reported for pristine undoped ZnO, HfO$_2$, TiO$_2$ and SnO$_2$ [1-8]. This has not only questioned the origin and nature of ferromagnetism in diluted wide band-gap semiconductors but also created new and exciting prospects for developing novel electronic components for applications like UV-sensors and spintronics devices. The observed ferromagnetism in pure ZnO is of particular interest due to the variety of potential applications that can be developed. Ferromagnetic order in pure ZnO has been attributed to defects in the lattice wherein both Zn and O vacancies have independently been claimed to generate the magnetization [2-4,9,10]. In a recent report Wang et al. [11] have presented a comprehensive theoretical investigation on vacancy induced magnetism in ZnO thin films using a generalized gradient approximation (GGA) in the well known density functional theory (DFT). The authors conclude that the observed ferromagnetism is due to Zn vacancies and not O vacancies as speculated earlier. The magnetic moment is believed to arise from the 2p unpaired electrons at O sites surrounding the Zn vacancy carrying a magnetic moment ranging from 0.490 to 0.740 μ_B.

All the experimental studies reported so far on the magnetism observed in the pristine nanoparticles/thin films have not focussed on determining the optimum conditions for the observed properties with respect to particle size (or film thickness), the appropriate conditions of the type of the ambient gas and its pressure. In this present study we report on the magnetic properties at RT for pure ZnO thin films deposited by DC Magnetron sputtering from a pure Zn metal target. After determining the optimum oxygen pressure for obtaining the highest magnetic properties in a thin film, a systematic study has been carried out to investigate the dependence of the magnetic properties on the film thickness in the range 100-1000 nm, all deposited under identical growth conditions and deposition parameters. We observe a unique phenomenon of a sequential transformation from ferromagnetism to the eventually expected diamagnetism as the film becomes thicker. Although earlier reports have shown ferromagnetic ordering in ZnO nano-particles, thin films, and nanorods [2-4,8], to the best of our knowledge, no study has been reported on the informative particle size/film thickness dependence of the evolution of magnetism in ZnO.

EXPERIMENTAL DETAILS

Thin films of ZnO (thickness ~100 - 1000 nm) were deposited on Al_2O_3 (0001) and Si (100) substrates at 350^0C under an oxygen partial pressure of 1.3×10^{-4} mBar from a pure Zn metal target (99.99 %, Kurt J.Lesker), by DC Magnetron Sputtering (Leybold-Heraeus). The chosen oxygen partial pressure was determined from the best magnetic property obtained for the 480 film. The deposited films were cooled down to room-temperature under controlled deposition atmosphere to preserve the oxygen related defects. The thickness of the films was determined using a dual beam SEM/FIB Nova 600 Nanolab (FEI Company). The crystallinity and the elemental mapping homogeneity test of the films were characterized respectively by XRD (Siemens D5000 powder X-ray diffractogram) and Energy Dispersive X-ray Spectrometer (OXFORD D7021) attached to a high resolution Hitachi 3000N SEM). The magnetic characterization has been carried out using the MPMS2-1 SQUID (Superconducting Quantum Interference Device from Quantum Design, USA) magnetometer.

RESULTS and DISCUSSION

From the structural characterization it is revealed that the ZnO films have grown prominently along the c-axis (002) direction where the angular position of the (002) peak varies from $33.85^0 - 34.33^0$ with increasing film thickness (see figure 1a). The positioning of the (002) peak at lower angles (2θ) indicates lattice expansion ($2\theta_{ZnO\ bulk}$ ~34.44^0) due to stress which may be due to a combination of (i) intrinsic defects (e.g. Zn_i and O_i) in the ZnO matrix and (ii) the strain at the film/substrate interface because of the lattice mismatch. We observe time dependent relaxation with long time annealing under controlled atmosphere [12,13]. We have studied the surface morphology of the films using an Atomic Force Microscope (Dimension 3100, Veeco) and find an almost linear increase in the surface roughness from about 3 to 12 nm as we increase the film thickness from ~100 – 720 nm. Figure 1b shows an example for the typical cross-section of the ~480 nm ZnO film deposited on Si substrate. Such thickness determination was carried out at various regions of the film with rather good reproducibility. Clearly, size effects are best studied in thin films using a FIB facility.

4

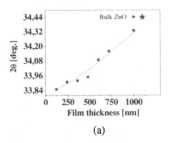 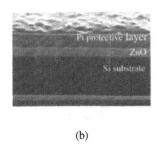

(a)   (b)

Figure 1. (a) Variation in (002) peak position with increasing film thickness (all films are deposited on Sapphire), the line is only a guide to the eye. (b) Cross-section of the 480 nm thick ZnO film using FIB showing a dense and porous free structure.

All the deposited ZnO films display RTFM with a characteristic magnetic saturation, low remanence and coercivity. Figure 2a shows a typical M(H) loop at RT for the 480 nm ZnO film. Figure 2b shows the obtained variation in M_S with increasing film thickness where the observed M_S values range from 0.18 – 0.62 emu/g (i.e. 0.005-0.018 μ_B/unit cell) with a maximum value found for the 480 nm thick film.

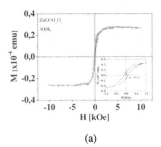 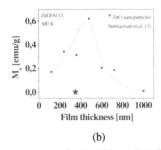

(a)   (b)

Figure 2. (a) Magnetic hysteresis loop at room-temperature for the 480 nm ZnO film (the data has been corrected for the diamagnetic background). Inset in figure 2a shows the hysteresis loop data at low fields indicating a coercive field of ~100 Oe. (b) Variation in M_S as a function of film thickness at room-temperature, the continuous line is a guide to the eye.

The observed M_S values are almost 3 orders higher than those reported earlier for ZnO powder, $1,8 \times 10^{-3}$ and nanoparticles $5,6 \times 10^{-4}$ emu/g [2,3] respectively. In thin films M_S has been reported in the range 0.2 – 35 emu/g [4,14] for films deposited by Pulsed Laser Deposition technique from a ceramic ZnO target. This clearly implies that the M_S values obtained in pristine ZnO depends not only on the method of preparation, but also on preparation condition and microstructure [15]. Clearly, data obtained on films under identical growth conditions are most reliable to systematically determine the film thickness dependence of the evolution of magnetism in ZnO. Such a study from nanoparticles alone would be subject to errors arising from average size, size distribution and the ill-defined morphology of the nanoparticles of larger diameter.

Also, note the much smaller M_S value (figure 2b) reported for nanoparticles which could be attributed due to the 'so-called' dead magnetic layers due to the spin disorder at the surface of the nanoparticle or other particle size dependent effects arising from the larger surface to volume ratio energetic for decreasing particle size.

We explain the observed transformation from ferromagnetism to diamagnetism with increasing film thickness as follows: The initial, almost linear, increase in M_S with increasing film thickness in the range 120-480 nm is due to the increasing number of cation vacancies in the films. A cation vacancy, due to the oxygen asymmetry around it is expected to develop a net spin which then interacts with another similar charged defect within an adequate range resulting in the overall long range ferromagnetic order [11]. The initial increase in the number of active interacting defects with film thickness results in the increasing net magnetization. In the thicker films (>480 nm) the relative ratio of uncoupled vs. coupled defects becomes larger since some defects are probably annealed out as suggested earlier. Hence, beyond the percolation limit the interplay between the coupled and neutral defects appears to decrease the overall magnetization and eventually (above 1 µm film thickness) we observe the well known diamagnetic property of bulk ZnO. Clearly, the details of the mechanism bringing about the observed magnetic transformations remains to be understood.

We now discuss the results of doping the pre-prepared 480 nm thick ZnO matrix having maximum M_S value with 1 at.% Mn (by co-sputtering the Zn metal target and a Mn metal target) keeping all other deposition parameters identical to those used in preparing the magnetic ZnO matrix. Figure 3 shows the XRD spectrum for the ZnO, $Zn_{0.99}Mn_{0.01}O$, a shift in the (002) peak position towards higher angle is observed on doping ZnO with 1 at.% Mn (from 33.94^0 to 34.0^0) that can be explained by the smaller atomic radius of Mn the atom (127 pm) in comparison to that of Zn the atom (134 pm). Such a shift on doping with Mn has been reported in literature with the while providing evidence for Mn occupying the Zn sites substitutionally in the ZnO lattice.

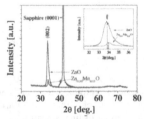

Figure 3. XRD spectrum for the ZnO and $Zn_{0.99}Mn_{0.01}O$ film. Inset shows an enlarged image of the spectrum at lower angles where the lattice compression upon Mn doping becomes more obvious.

From the SQUID data the net M_S at room-temperature in the 480 nm thin ZnO film is found to increase by almost ~26% (from 0.62 emu/g to 0.78 emu/g) on doping with ~1 at.% Mn (with a Mn contribution of 0,23 μ_B/Mn). Whereas, on increasing the Mn conc. to 2 at.% the net M_S decreases with ~62% to 0.24 emu/g (see figure 4). These calculations assume homogeneous distribution of the dopant atoms in the ZnO lattice and clearly show that the maximum value for M_s is obtained at 1 at.% Mn in the optimized matrix.

6

Ferromagnetic ordering in transition metal, TM, doped ZnO is believed to arise from ferromagnetic coupling between the un-paired 3d electrons of the TM as a result of the hybridization of the 2p states of O with the 3d state of the TM [16-22]. It is understood that we in the present case have a ferromagnetic diffuse defect band that is carrying the magnetization between localized moments, however the difference in the electronic structure on Mn doping has yet to be understood in-order to explain the change in magnetization with the dopant concentration in the Mn doped ZnO.

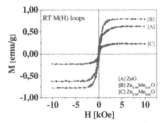

Figure 4. Variation in net magnetization at room-temperature in doped ZnO films as a function of Mn concentration

It may be useful to point out that in all our films the temperature dependent magnetization data (not shown here) at low fields (~100 Oe) for the doped ZnO films do not indicate any other possible additional impurity phases and the observed magnetic properties are intrinsic to $Zn_{1-x}Mn_xO$. Furthermore, our XPS studies on these films show strong hybridization between the O 2p and Mn 3d lines stating that the net moment to be a combination of the intrinsic defects in the lattice and the ferromagnetic coupling between the un-paired 3d electrons of the TM using the defect band as a mediator (*to be published*).

CONCLUSIONS

In summary, we have carried out a systematic study on the thickness dependency of defect induced magnetism in pure ZnO thin films (ranging 120-1000 nm) and find an optimum thickness of 480 nm for the present growth conditions to give a maximum saturation magnetization of ~0.62 emu/g (0,018 μ_B/unit cell). We have also shown systematically (under identical growth conditions) how one can tune the ferromagnetic matrix of ZnO by introducing TM atoms into the host matrix. Highest M_S value for an optimum film thickness is observed for 1 at.% Mn doping (0.78 emu/g), but decreases at higher Mn conc. The effective moment per Mn atom calculates to be 0,23 μ_B/Mn when doping the ZnO with 1 at.% Mn. The low moment per Mn atom as compared to the theoretical value (5 μ_B/Mn) could partly be due to the fact that not all the Mn atoms are contributing to the long range magnetic ordering.

The net magnetic moment is found to be a combination of the moment induced by (i) intrinsic defects (i.e. from the 2p unpaired electrons at O sites surrounding the vacancy at cation site) and (ii) ferromagnetic coupling between the un-paired 3d electrons of the TM as a result of the hybridization of the 2p states of O with the 3d state of the TM. The change in the electronic properties of ZnO due to the dopant is an additional factor to be elucidated in understanding the overall defect induced mechanism for ferromagnetism in the doped ZnO.

ACKNOWLEDGMENTS

We gratefully acknowledge the intense discussions with Prof. Gillian Gehring from Sheffield University. This work is supported by the Swedish Funding Agency VINNOVA, Swedish Research Council, and the Hero-M Excellence Centre at KTH. XJ acknowledges a post doctoral fellowship from Carl Trygger's Foundation in Sweden.

REFERENCES

1. N. H. Hong, J. Sakai, N. Poirot and V. Brizé, Phys. Rev. B **73** (2006) 132404
2. S. Banerjee, M. Mandal, N. Gayathri and M. Sardar, Appl. Phys. Lett. **91** (2007) 182501
3. A. Sundaresan, R. Bhargavi, N. Rangarajan, U. Siddesh and C.N.R. Rao, Phys. Rev. B **74** (2006) 161306
4. Q. Xu, H. Schmidt, S. Zhou, K. Potzger, M. Helm, H. Hochmuth, M. Lorenz, A. Setzer, P. Esquinazi, C. Meinecke and M. Grundmann, Appl. Phys. Lett. **92** (2008) 082508
5. M.Venkatesan, C. B. Fitzgerald and J.M.D. Coey, Nature (London) **430** (2004) 630
6. J. M. D. Coey, M. Venkatesan, P. Stamenov, C. B. Fitzegerald and L.S. Dorneles, Phys. Rev. B **72** (2005) 024450
7. Z. Yan, Y. Ma and D. Wang, J. Wang, Z. Gao, L. Wang, P. Yu and T. Song, Appl. Phys. Lett. **92** (2008) 081911
8. S. Kumar, Y.J. Kim, B.H. Koo, S. Gautam, K. H. Chae, R. Kumar and C.G. Lee, Materials Letters **63** (2009) 194-196
9. W. Yan, Z. Sun, Q. Liu, Z. Li, Z. Pan, J. Wang, S. Wei, D. Wang, Y. Zhou and X. Zhang, Appl. Phys. Lett. **91** (2006) 062113
10. Q. Xu, H. Schmidt, L. Hartman, H. Hochmuth, M. Lorenz, A. Setzer, P. Esquinazi, C. Meinecke and M. Grudman, Appl. Phys. Lett. **91** (2007) 092503
11. Q. Wang, Q. Sun, G. Cheng, Y. Kawazoe and P. Jena, Phys. Rev. B **77** (2008) 205411
12. P. Erhart and K. Albe, Appl. Phys. Lett. **88** (2006) 201918
13. Y. Bing-Chu, L. Xiao-Yan, G. Fei and M. Xue-Long, J. Cent. South Univ. Technol. **15** (2008) 449-453
14. N. H. Hong, J. Sakai and V. Brizé, J. Phys.: Condens. Matter **19** (2007) 036219
15 M. Kapilashrami, R. V. Upadhyay, V. Ström, L. Belova and K. V. Rao, AIP Conference Proceedings, Vol. **1003** (2008) 255-257
16. L. M. Huang, A. L. Rosa and R. Ahuja, Phys. Rev. B **74** (2006) 075206
17. J-H. Guo, A. Gupta, P. Sharma, K. V. Rao, M. A. Marcus, C. L. Dong, J. M. O. Guillen, S. M. Butorin, M. Mattesini, P. A. Glans, K. E. Smith, C. L. Chang and R. Ahuja, J. Phys. Condens. Matter **19** (2007) 172202
18. P. Thakur, K. H. Chae, J.Y. Kim, M. Subramanian, R. Jayavel and K. Asokan, Appl. Phys. Lett. **91** (2007) 162503
19. T. Dietl, J. Phys.: Condens. Matter **19** (2007) 165204
20. P. Sharma, A. Gupta, K. V. Rao, F. J. Owens, R. Sharma, R. Ahuja, J. M. Osorio-Guillen, B. Johansson and G. A. Gehring, Nature Materials **2** (2003) 673
21. C. L. Dong, C. Persson, L. Vayssieres, A. Augustsson, T. Schmitt, M. Mattesini, R. Ahuja, C. L. Chang and J-H. Gou, Phys. Rev B **70** (2004) 195325
22. Z. Fu-Chun, Z. Zhi-Yong, Z. Wei-Hu, Y. Jun-Feng and Y. Jiang-Ni, Chin. Phys. Lett. Vol. **26** No.1 (2009) 016105

Mater. Res. Soc. Symp. Proc. Vol. 1183 © 2009 Materials Research Society   1183-FF01-11

Formation of Cr-Rich Nano-Clusters and Columns in (Zn,Cr)Te Grown by MBE

Yôtarô Nishio[1], Kôichirô Ishikawa[1], Shinji Kuroda[1], Masanori Mitome[2], Yoshio Bando[2]
[1]Institute of Materials Science, University of Tsukuba, 1-1-1 Tennoudai, Tsukuba, 305-8573, Japan
[2]Advanced Materials and Nanomaterials Laboratories, National Institute for Materials Science, 1-1 Namiki, Tsukuba, 305-0044, Japan

ABSTRACT

The correlation between the Cr aggregation and magnetic properties are investigated for the series of $Zn_{1-x}Cr_xTe$ films grown by MBE with a systematic variation of growth conditions. Structural and chemical analyses using TEM and energy-dispersive X-ray spectroscopy (EDS) reveal that the crystallinity and the Cr distribution change significantly with the substrate temperature during the MBE growth. For a relatively low average Cr content $x \cong 0.05$, it is found that the crystal quality is improved with the increase of the substrate temperature. For a higher average Cr content $x \cong 0.2$, the shape of Cr-rich regions is transformed from isolated clusters into one-dimensional nanocolumns with the increase of the substrate temperature. The direction of the nanocolumn formation changes depending on the crystallographic orientation of the grown films. In the magnetization measurements, anisotropic magnetic properties are observed in the films in which Cr-rich nanocolumns are formed in the vertical direction, depending on the relation between the direction of the nanocolumns and the applied magnetic fields.

INTRODUCTION

Search for novel semiconducting materials exhibiting ferromagnetism with a high transition temperature is one of the most challenging topics in today's materials science and technology. So far, a broad class of diluted magnetic semiconductors (DMSs) have been investigated and some of them have been claimed to be room-temperature ferromagnetic semiconductors [1,2]. However, the intrinsic nature of ferromagnetism has sometimes been controversial, both positive and negative experimental results being reported even for the same materials [3]. Recently, it has become realized that the distribution of magnetic elements in the host crystal is a key dominating the magnetic properties of DMSs [4]. In some of DMSs, the solubility limit of magnetic elements in the host semiconductor is quite low and the phase separation into regions with low and high contents of magnetic elements is expected beyond this limit [5]. Indeed, it has been reported in various DMSs [6-10] that nano-scale regions containing high-content magnetic elements are formed in the crystals and the ferromagnetic properties are accordingly enhanced. In our recent publication on (Zn,Cr)Te [9], it has been demonstrated that the Cr distribution can be controlled in a systematic way by the co-doping of donor or acceptor impurities or by the ratio of flux supplies during the MBE growth; the distribution of Cr ions becomes inhomogeneous and Cr-rich regions are formed coherently in the crystals co-doped with iodine (I) as a donor impurity or grown with a surplus supply of Zn flux. As an explanation for the observed difference in the uniformity of Cr distribution, it has been proposed [11] that the aggregation energy between Cr ions is manipulated through the shift of Fermi energy due to the co-doping of donor or acceptor impurities or the deviation from stoichiometry under different Zn/Te flux supply ratio. In the case of inhomogeneous Cr distribution, the Cr-rich regions formed in the crystal act as ferromagnetic clusters and the whole crystal exhibits superparamagnetic properties due to the magnetic anisotropy of the respective clusters [12]. At the same time, this magnetic anisotropy gives the appearance of high-temperature ferromagnetism, such as hysteretic behaviors in the magnetization curve.

In the present study, we have investigated the structural, chemical and magnetic properties of the series of I-doped (Zn,Cr)Te films grown with a systematic variation of growth conditions such as the substrate temperature during the MBE growth, the growth rate, and the crystallographic orientation of the substrate. Combined analyses of the Cr distribution and the magnetic properties using the spatially resolved energy-dispersive X-ray spectroscopy (EDS) in the TEM observation and the SQUID magnetometry respectively reveal that the crystallinity and the shape of Cr-rich regions change significantly with the substrate temperature during the growth.

EXPERIMENTAL PROCEDURE

Thin films of $Zn_{1-x}Cr_xTe$ were grown by solid-source molecular beam epitaxy (MBE) on GaAs (001) or (111) substrates. For the co-doping of donor impurity, an additional flux of iodine (I) was supplied from a compound source of CdI_2. We have prepared two series of I-doped $Zn_{1-x}Cr_xTe$ films with relatively low and high average Cr contents of $x \cong 0.05$ and 0.2, which were grown respectively with a systematic variation of growth conditions such as the substrate temperature during the growth, the growth rate, and the crystallographic orientation of the substrate. High-resolution TEM analysis was performed for cross-sectional pieces (thickness ~ 100nm) cut from the grown films using focused ion beam (FIB). The composition of constituent elements was estimated using EDS and the nano-scale spatial distribution of Cr was examined by mapping the intensity of the Cr K_α line in the scanning TEM mode. The magnetization of the films was measured using SQUID magnetometer in the reciprocating sample measurement mode with magnetic fields perpendicular or parallel to the film plane. From the dependences of magnetization on temperature and magnetic field, characteristic features of magnetic properties of the grown films were derived.

EXPERIMENTAL RESULTS AND DISCUSSION

We have grown I-doped $Zn_{1-x}Cr_xTe$ films with relatively low and high value of the average Cr content $x \cong 0.05$ and 0.2 with a systematic variation of growth conditions such as the substrate temperature during the growth, the growth rate, and the crystallographic orientation of the substrate. In this article, we mainly focus on the variation with the substrate temperature. For a low average Cr content $x \cong 0.05$, we have prepared a series of the films grown on the (001) surface at different substrate temperatures in the range of $T_S = 240 \sim 390°C$. Figure 1 summarizes the results of TEM and EDS observation of the series of the films. In the TEM lattice and diffraction images shown in the upper row of Fig. 1, the crystal quality changes with the substrate temperature T_S; at an intermediate substrate temperature $T_S = 300°C$, which is a typical value for the growth of the host binary compound ZnTe, there appears some regions containing stacking faults along the {111} planes in the lattice image, though the crystal structure exhibits

10

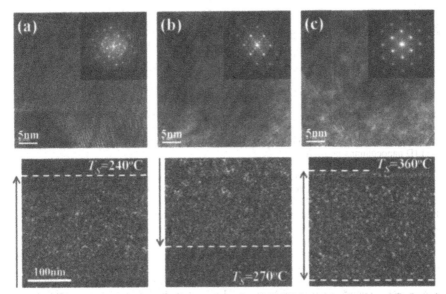

Figure 1. The lattice and diffraction images (upper) and the EDS mapping images of Cr (lower) of I-doped $Zn_{1-x}Cr_xTe:I$ films ($x \cong 0.05$) grown on the (001) surface at a substrate temperature of (a) $T_S = 240°C$, (b) $T_S = 270°C$, (c) $T_S = 360°C$, respectively. In the EDS mapping images, the extent and the boundary of $Zn_{1-x}Cr_xTe:I$ layer are indicated by black arrows and white dashed lines, respectively.

dominantly zinc-blende (ZB) structure. Corresponding to these stacking faults, additional spots appear at the one-third positions of the main hexagonally-arranged spots of ZB structure in the diffraction image. At the lowest substrate temperature of $T_S = 240°C$, the crystal quality is deteriorated; the diffraction image exhibits a complicated pattern having polycrystal-like features in addition to the stacking faults. On the other hand, at a higher substrate temperature of $T_S = 360°C$, the crystal quality is significantly improved; there appear scarcely any stacking faults in the lattice image and the diffraction image exhibits almost perfect hexagonal pattern of the ZB structure. The EDS mapping images of Cr in the respective films are shown in the lower row of Fig. 1. Though the Cr distribution is inhomogeneous in all the films, the degree of inhomogeneity is reduced slightly with the increase of T_S.

The magnetization measurements of these films reveal that the magnetic properties are correlated with the crystal quality and the Cr distribution. The apparent ferromagnetic transition temperature $T_C^{(app)}$ is deduced from Arrott plot analysis of the magnetization vs. magnetic field (M-H) curve. The film grown at the lowest substrate temperature of $T_S = 240°C$ does not exhibit ferromagnetic transition even at the lowest temperature 2K. This is considered due to the deterioration of crystal quality. On the other hand, in the range of higher substrate temperatures, $T_C^{(app)}$ exhibits a tendency to decrease gradually with the increase of T_S; from $T_C^{(app)} = 310K$ at $T_S = 270°C$ to $T_C^{(app)} = 240K$ at $T_S = 390°C$. This decreasing tendency of $T_C^{(app)}$ is considered to be correlated with a reduced degree of the inhomogeneity of Cr distribution with T_S.

11

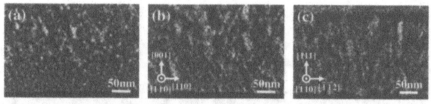

Figure 2. The EDS mapping images of Cr in $Zn_{1-x}Cr_xTe:I$ films ($x \cong 0.2$) grown (a) at a substrate temperature $T_S = 300°C$ on the (001) surface, (b) at $T_S = 360°C$ on the (001) surface, and (c) at $T_S = 360°C$ on the (111) surface. The shape of Cr-rich regions is 0D clusters in (a) and 1D nanocolumns in (b), (c).

For a higher average Cr content $x \cong 0.2$, we have grown I-doped $Zn_{1-x}Cr_xTe$ films at an intermediate and high substrate temperatures of $T_S = 300°C$ and $360°C$ on the (001) and (111) surfaces, respectively. Figure 2 (a), (b) compares the Cr mapping images of the films grown on the (001) surface at the different substrate temperatures $T_S = 300°C$ and $360°C$. It is clearly demonstrated that the shape of Cr-rich regions changes with the increase of T_S; Cr-rich regions are formed as isolated clusters at an intermediate temperature $T_S = 300°C$, while they take the form of stripes at a higher temperature $T_S = 360°C$. In the latter case, Cr-rich stripes with ~10 nm width are arranged with an interval of 10~20nm. This image indicates that Cr-rich regions are formed as one-dimensional (1D) columns. These stripes are slanted against the growth direction with an angle around 55 degree, which suggests that Cr atoms have the tendency to aggregate along the {111} plane of the ZB structure. The difference in shape of the Cr-rich regions by the substrate temperature can be interpreted as a result of a different degree of migration of the impinging atoms on the growing surface. According to a theoretical simulation [13], the spinodal decomposition in the layer-by-layer growth mode results in the formation of 1D columnar regions containing magnetic impurity in rich contents. It is considered that the enhanced migration on the growing surface at an elevated temperature promote the aggregation of Cr atoms in such areas where Cr-rich areas are already formed in the layer just below the surface.

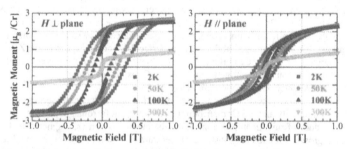

Figure 3. Magnetization vs. magnetic field (*M-H*) curves of a $Zn_{1-x}Cr_xTe:I$ ($x = 0.22$) film grown on the (111) surface. (the same film as shown in Fig. 2 (c)). The left (right) panel shows the result under magnetic fields perpendicular (parallel) to the film plane. The coercive field H_C under perpendicular fields ($H_C = 0.33T$ at 2K) is much larger than under parallel fields ($H_C = 0.14T$ at 2K)

As a result, Cr-rich areas in the respective layers are piled up and Cr-rich regions form themselves into 1D columns. When a (Zn,Cr)Te film are grown on the surface with a different crystallographic orientation, it is found that the direction of the Cr-rich nanocolumns changes. Figure 2 (c) shows the Cr mapping image of a film grown on the (111) surface at $T_S = 360°C$. Cr-rich nanocolums are formed in an almost vertical direction against the surface. This is also consistent with the tendency that Cr atoms aggregate along the {111} plane of the ZB structure.

The magnetization measurements reveal that the magnetic properties are different according to the shape of Cr-rich regions formed in the crystals. In comparison between the two (001) films grown at $T_S = 300°C$ and $360°C$ (Fig. 2 (a), (b)), the ferromagnetic transition temperature $T_C^{(app)}$ is not different so much, but the superparamagnetic features are more pronounced in the film grown at $T_S = 360°C$; the blocking temperature T_B, which is determined as a peak position in the magnetization vs. temperature (M-T) curve measured in the zero-field-cooled process, is higher in the film grown at $T_S = 360°C$ than $T_S = 300°C$. This is attributed to a difference in the energy barrier due to the magnetic anisotropy between Cr-rich clusters and nanocolumns [12]; the energy barrier should be larger for the nanocolumns with a larger volume than the clusters. In addition, the anisotropic magnetization depending on the relation between the magnetic field direction and the film plane is observed in the (111) films, in which Cr-rich nanocolumns are formed in the almost vertical direction. Figure 3 compares the magnetization vs. magnetic field (M-H) curves of the (111) film grown at $T_S = 360°C$ (Fig. 2 (c)) under magnetic fields perpendicular and parallel to the film plane, respectively. As seen in the figure, hysteretic behaviors are more pronounced with larger coercive fields under perpendicular magnetic fields than parallel fields. This anisotropy suggests that the easy magnetization axis is along the axial direction of the nanocolumns. This is quite reasonable considering that the demagnetization field in this columnal shape is larger under magnetic fields along the axial direction.

CONCLUSIONS

We have investigated the dependences of the structural, chemical and magnetic properties of $Zn_{1-x}Cr_xTe$ films on the substrate temperature during MBE growth for low and high average contents of $x \cong 0.05$ and 0.2. For a relatively low average Cr content $x \cong 0.05$, the crystal quality is improved with the reduction of stacking faults with the increase of the substrate temperature. For a higher average Cr content $x \cong 0.2$, the shape of Cr-rich regions is transformed from isolated clusters at an intermediate substrate temperature $T_S = 300°C$ into one-dimensional nanocolumns at a high temperature of $T_S = 360°C$. In addition, it is found that the direction of the nanocolumn formation changes according as the films are grown on the (001) or (111) surface, suggesting that Cr atoms tend to aggregate along the {111} plane of zinc-blende structure. In the magnetization measurements, anisotropic properties are observed in the films containing Cr-rich nanocolumns, originating from the relation between the direction of the nanocolumns and the applied magnetic fields.

13

ACKNOWLEDGMENTS

This work has partially been supported by the Grant-in-Aids for Scientific Research (Basic Research (B) and Priority Areas) and by "Nanotechnology Support Project" of the Ministry of Education, Culture, Sports, Science and Technology (MEXT), Japan.

REFERENCES

1. C. Liu, F. Yun and H. C. Morkoç, *J. Mater. Sci.: Mater. Electron.* **16**, 555 (2005).
2. S. A. Chambers, *Sur. Sci. Rep.* **61**, 345 (2006).
3. J. M. D. Coey, *Curr. Opinion Solid State Mater. Sci.* **10**, 83 (2006).
4. T. Dietl, in *Spintronics*, edited by T. Dietl, D. D. Awschalom, M. Kamińska and H. Ohno (*Semiconductors and Semimetals* **82**, Academic Press, 2008), pp. 371-432; H. Katayama-Yoshida, K. Sato, T. Fukushima, M. Toyoda, H. Kizaki and A. v. Dinh, ibid, pp.433-454.
5. K. Sato, H. Katayama-Yoshida and P. H. Dederichs, *Jpn. J. Appl. Phys.* **44**, L948 (2005).
6. L. Gu *et al.*, *J. Magn. Magn. Mater.* **290-291**, 1395 (2005).
7. M. Jamet *et al.*, *Nature Mater.* **5**, 653 (2006).
8. D. Bougeard, S. Ahlers, A. Trampert, N. Sircar and G. Abstreiter, *Phys. Rev. Lett.* **97**, 237202 (2006).
9. S. Kuroda, N. Nishizawa, K. Takita, M. Mitome, Y. Bando, K. Osuch and T. Dietl, *Nature Mater.* **6**, 440 (2007).
10. A. Bonanni *et al.*, *Phys. Rev. Lett.* **101**, 135502 (2008).
11. T. Dietl, *Nature Mater.* **5**, 673 (2006).
12. K. Sato, T. Fukushima, H. Katayama-Yoshida, *Jpn. J. Appl. Phys.* **46**, L682 (2007).
13. T. Fukushima, K. Sato, H. Katayama-Yoshida, and P. H. Dederichs, *Jpn. J. Appl. Phys.* **45**, L416 (2006).

Magnetic Resistive RAM

Mater. Res. Soc. Symp. Proc. Vol. 1160 © 2009 Materials Research Society   1160-H03-04-FF03-04

Novel Magnetoresistive Structures Using Self-Assembly and Nanowires on Si

Mazin Maqableh*, Xiaobo Huang and Bethanie J. H. Stadler
Department of Electrical and Computer Engineering, University of Minnesota, Minneapolis, Minnesota 554553

ABSTRACT

Anodic Aluminum Oxide (AAO) was grown both as free-standing membranes and as integrated layers on Si as templates for arrays of magnetoresistive nanowires. The barrier layer was completely removed in both cases and Co/Cu multilayered nanowires were successfully grown in these templates by DC electrodeposition. Magnetic hysteresis loops and current-perpendicular-to-plane giant magnetoresistance (CPP-GMR) up to 25% were measured for nanowires grown in free standing AAO templates and in templates grown on Si. Spin transfer torque (STT) switching was also measured for multilayers grown in free standing templates with a switching current density of 2.7×10^8 A/cm^2.

INTRODUCTION

Anodic Aluminum Oxide (AAO) is a promising template material for fabricating nanowires because of its self-assembled nanopores whose dimensions can be precisely controlled by tuning the different anodization parameters [1]. As well as having free standing AAO templates, the AAO can be integrated onto Si substrate [2-5] to open the road of making devices such as MRAM and catalysts that benefit from the combination of silicon processing and the self assembly properties of AAO.

AAO templates can be made using two-step anodization which results in highly ordered and straight nanopores [6]. A major concern for integrated nanowires is the removal of the barrier layer, which is a thin aluminum oxide layer existing at the bottom of the pores. This must be removed or thinned before efficient electrochemical deposition of nanowires can occur. Several methods have been used to remove this barrier. One method involves pore widening by phosphoric or chromic acid which will result in removal of the barrier layer as well as widening the pores [3, 5]. This method is disadvantageous in the sense that pore size is not preserved. Another method uses Ar ion-milling to break the barrier layer [7]. This method has two disadvantages. It requires an ultra thin AAO template so that Ar ions can reach the bottom of the pores with sufficient energy to break the barrier layer. It also damages the surface of the AAO as well as etching it, so the AAO thickness in this method is not preserved. A third method is to perform the second anodization for a very long time. A spike in the time dependence current curve during this step is used as a sign to manually stop the anodization process [3, 4].

Metallic nanowires can be grown in these templates by DC [3, 4, 8] and AC [9, 10] electrochemical deposition. Co/Cu multilayered nanowires have also been electrodeposited in free standing AAO templates using a mixture electrolyte that contains both Co and Cu cations [8, 12-14]. These electrochemically deposited multilayered nanowires have shown current-perpendicular-to-plane giant magnetoresistance (GMR) [8, 11] as well as a spin transfer torque phenomenon (STT) [12].

In this work, anodic aluminum oxide (AAO) was grown as free standing templates and also as successfully integrated templates on Si. The three barrier removal methods described above were tested here. Only the third method worked in the complete removal of the barrier layer which was further investigated by electrodepositing Cu into pores after attempting barrier removal. Therefore, this method was used in all the subsequent work presented in this paper. Co/Cu multilayered nanowires were successfully grown and their magnetic properties such as MH loops, giant magnetoresistance (GMR) have been measured. Spin transfer torque switching was also measured in the nanowires that were grown in free standing AAO templates.

EXPERIMENTAL

Free standing AAO templates

Two-step anodization [6] was used to make AAO templates from electropolished foils of Al metal. After anodization, the remaining Al metal was etched away with mercuric chloride, leaving oxide templates that contained nanopores with diamters 10 nm to 70 nm (using sulphuric acid, H_2SO_4), or from 40 nm to 150 nm (using oxalic acid, $H_2C_2O_4$). The barrier layers were etched by floating the templates on a mixture of phosphoric and chromic acid, and Cu films were sputtered onto the back of the templates.

AAO on Si

A 1µm aluminum film was evaporated using e-beam evaporation on a silicon substrate coated with titanium and copper films (200nm each). Two-step anodization process at 18C in 0.4M oxalic was then used to make a 600-700nm anodic aluminum oxide (AAO). The anodization voltage was kept constant at 40V during the two anodization steps. After the first anodization which was run for 4 minutes, the resulting AAO was etched away using a mixture of 1.8wt% chromic acid and 6wt% phosphoric acid for 30-45 minutes at 60C. The resulting aluminum, which was about 600nm thick, was second anodized using the same parameters to create a 600-700nm thick AAO with pore diameter of 40nm and inter-pore spacing of about 100nm. In addition to growing these latter pores directly onto Si, they were also grown onto Co(20nm)/Cu(10nm)/Co(10nm) thin films that were evaporated onto Si.

The barrier layer was removed by running the second anodization step for a much longer time. For the case of Si/Ti/Al substrate, which was used initially, this method failed in the removal of the barrier. However, for the case of Si/Ti/Cu/Al substrate, this method succeeded in removing this barrier.

Magnetorresistive nanowires.

DC electrochemical deposition was carried out at room temperature to grow Co/Cu multilayered nanowires in the membrane. The electrolyte solution was made of 155 g/L $CoSO_4.7H_2O$, 1.13 g/L $CuSO_4$ and 50g/L HBO_3. Cyclic voltametry was used to determine the cathode potential for Cu and Co deposition (-0.52 and -1 volts respectively) [13]. The purpose of HBO_3 was to maintain the pH value of the solution at 3.7.

Multilayered Co/Cu nanowires were fabricated with different layer thicknesses by controlling the deposition time of each layer. For the Si-integrated nanowires, 50 bi-layers of

Co(7.5nm)/Cu(5nm) were grown with a thick Cu layer (about 50nm) deposited prior to and after the deposition of the multilayered nanowires. Field emission scanning electron microscope (FESEM) was used to study the structure of the AAO and nanowires. Magnetic properties of the samples were verified by a vibrating sample magnetometer (VSM). MR was measured using an ac and dc magnetotransport systems with a bias current of 1mA. STT switching was measured in multilayered nanowires grown in the free standing AAO membranes.

RESULTS AND DISCUSSION

Free standing AAO templates

As the AAO was formed, using a two-step anodization process, columnar nanopores self-assembled inside the oxide to form a close-packed array. The pore diameters were varied from 10-60nm by changing the anodization conditions [1, 14]. As the diameter of the AAO nanopores decreased, the distance between the nanopores also decreased. The free-standing membranes had pores with lengths of 17μm.

AAO on Si

It was possible to vary nanopores diameter without changing spacing. This is a great way to analyze the affect of interwire magnetic interactions on the magnetoresistive (MR) properties, which will be done in the future. The initial MR characterization is reported. Figure 1 shows SEM images of self-assembled nanopores grown on Si with different pore sizes and fixed spacing.

Fig. 1. SEM images of self-assembled nanopores grown on Si (a) top view with pore size of 40nm and 100nm spacing (0.4M oxalic acid at room temperature), (b) top view with pore size of 30nm and 100nm spacing (0.3M oxalic acid at 5C), and (c) cross-sectional view of AAO on Si/Ti with a barrier layer.

Figure 2 shows the time dependence of the current during the long second anodization step. Fig. 2-a represents the second anodization of Al on Si/Ti substrate where the very long second anodization did not result in any spikes in the current, which means the barrier was not broken. Fig. 2-b represents the case of having Al on Si/Ti/Cu substrate where the current increased very rapidly at about 270s, and then came back to its low value at about 350s. The flat part on the top is from the instrument current limit which is 105mA. During this spike, the Cu film tended to be very reactive to the acid and it was completely gone after this spike. Thereafter, the sample behaved much likely as the Si/Ti/Al substrate. We thus used this spike as a sign to

19

manually stop the anodization process as shown in Fig. 2-c. Manually stopping the anodization process at this point of the curve was proven to be an ideal method to completely get rid of the barrier layer. However, if this anodization step was not given the sufficient time for the rapid increase in current to show up as shown in Fig. 2-d, the barrier layer remained at the bottom of the pores.

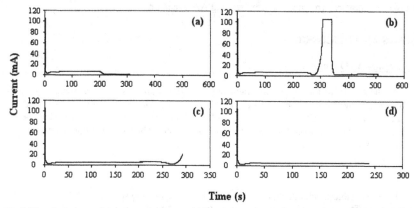

Fig. 2. Time dependence of the current during the (a) 2nd anodization of Al on Si/Ti substrate for 500s, (b) 2nd anodization of Al on Si/Ti/Cu substrate for 500s, (c) manually stopped 2nd anodization of Al on Si/Ti/Cu substrate at the early stages of the spike, and (d) manually stopped 2nd anodization of Al on Si/Ti/Cu substrate before the spike.

Attempts at electrochemical deposition inside the pores was used as evidence of the presence or the absence of a barrier layer in each sample. Figure 3 shows the behavior of the current during the electrodeposition of Cu in pores of samples whose barrier layers were treated using the three methods described above.

As seen in Fig. 3, only the third method resulted in a large, uniform current during deposition that can only be obtained when a good contact was made between the electrolyte and the metal electrode below the pores. The remaining barrier layer in samples processed by the first two methods essentially acts as an insulator that prevents the electrolyte from accessing the electrode.

Magnetoresistive nanowires

Using the free standing AAO templates, the highest magnetoresistance was found in nanowires that had hysteresis loops that were identical as measured in plane and perpendicular to the plane. The highest measured MR (= $\Delta R/R$ = 11%) of the multilayers was calculated as 33% by subtracting the resistance of the Cu leads on either side of the multilayers from the denominator [14]. Figure 4-a shows MH loops of a 50 bilayers of Co(7.5nm)/Cu(5nm), in which the sample appears to have an easy axis perpendicular to the nanowires.

It was found that multilayered Co/Cu nanowires grown in free standing AAO templates had similar magnetoresistance (MR) behavior as comparable nanowires grown on Si. However, compared to the relatively high MR value obtained in nanowires grown in free standing templates, nanowires on Si had a lower value which was measured to be 2-3%, Fig. 4-b,c. This may be due to larger lead resistance, which is difficult to measure with these integrated samples, but further measurements are underway. The MR curve had a broader peak for the case where the field was applied parallel to the wires because the demagnetization fields due to the shape anisotropy of the Co layers inhibit switching until higher applied fields.

Spin transfer torque (STT) was measured in the nanowires that were inside free standing AAO. For 10-60nm diameter nanowires, the change in resistance due to STT was around 6% which represents the full magnetoresistance of the larger wires [14] but only half that of the smaller nanowires. It is therefore concluded that the 10-nm Co layers do not align antiparallel to parallel as fully at the switching current density of $J_{AP-P} = 2.7 \times 10^8$ A/cm^2 compared to the larger wires which switch at $J_{AP-P} = 3.2 \times 10^7$ A/cm^2. Au nanocontacts were electroplated into nanopores on Co(20nm)/Cu(10nm)/Co(10nm) thin films evaporated onto Si to prove the feasibility of using this technique in a wide range of configurations to study point-contact magnetoresistance and/or microwave response.

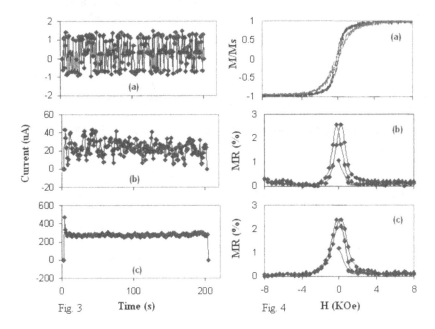

Fig. 3   Time (s)   Fig. 4   H (KOe)

Fig. 3. Time dependence of the current during DC electrodeposition of Cu in pores of samples whose barrier layers were treated using (a) Ar ion milling for 15mins, (b) pore widening in 5wt% H$_3$PO$_4$ for 45mins, and (c) very long second anodization.

Fig.4. MH loops of Co(7.5nm)/Cu(5nm) multilayered (50 bi-layers) nanowires with a diameter of 40nm when the field is perpendicular (blue) and parallel (pink) to the wires (a), and magnetoresistance when the field is perpendicular (b) and parallel (c) to the wires.

CONCLUSIONS

Anodic aluminum oxide (AAO) was successfully integrated as free standing membranes and as successfully integrated templates on Si with pore sizes that were varied between 10-60 nm by adjusting the anodization parameters. Into these templates, Co/Cu nanowires were successfully fabricated by DC electrodeposition with an easy axis perpendicular to the wires length. These mutlilayers showed a current-perpendicular-to-plane giant magnetoresistance (CPP-GMR) as well as spin transfer torque (STT) switching. Co/Cu multilayers grown on Si had a GMR ratio of 2-3% which is lower than that of the same multilayers grown in free standing templates. This might be due to the lead resistance in this integrated structure which is currently under study. Co/Cu multilayered nanowires grown in 10nm free standing AAO templates had a spin transfer torque (STT) phenomenon with a switching current density of 2.7×10^8 A/cm^2. Multilayers grown on Si are currently under construction and this switching phenomenon will be measured in the future. With diameters in the 10-60 nm range and integration with Si, these nanostructures have great potential for future nanosensors, MRAM and microwave oscillator arrays.

REFERENCES

1. A. Belwalkar, E. Grasing, W. Van Geertruyden, Z. Huang, W.Z. Misiolek, J. Membr. Sci., **319**, 192–198 (2008).
2. Bo Yan, Hoa T. M. Pham, Yue Ma, Yan Zhuang, Pasqualina M. Sarro, Appl. Phys. Lett. **91**, 053117 (2007).
3. S.K. Lim, G.H. Jeong, I.S. Park, S.M. Na, S.J. Suh, J. Magn. Magn. Mater. **310**, e841-e842 (2007).
4. Cai-Ling Xu, Hua Li, Guang-Yu Zhao, Hu-Lin Li, Appl. Surf. Sci. **253**, 1399-1403 (2006).
5. Filimon Zacharatos, Violetta Gianneta and Androula G Nassiopoulou, Nanotechnology **19**, 495306 (2008).
6. H. Masuda and K. Fukuda, Science **268**, No. 5216, 1466-1468 (1995).
7. M. Tian, S. Xu, J. Wang, N. Kumar, E.Wertz, Q. Li, P.M. Campbell, M. Chan, and T.E. Mallouk, *Nano Lett.* **5**, 697-703 (2005).
8. K. Liu, K. Nagodawithana, , P.C Searson, C.L Chien, Phys. Rev. B **51**, no. 11, 7381-4 (1995).
9. M. Kashi, A. Ramazani and A. Khayyatian, J. Phys. D: Appl. Phys. **39**, 4130–4135 (2006).
10. Jinxia Xu, Yi Xu, Mater. Lett. **60**, 2069–2072 (2006).
11. L. Gravier, J.-E Wegrowe, T. Wade, A. Fabian, and J.-P. Ansermet, IEEE Trans. Magn. **38**, no. 5, 2700-2702 (2002).
12. T. Blon,a_ M. Matefi-Tempfli, S. Matefi-Tempfli, L. Piraux, S. Fusil, R. Guillemet, K. Bouzehouane, C. Deranlot, and V. Cros, J. Appl. Phys. **102**, 103906 (2007).
13. Liwen Tan and Bethanie J. H. Stadler, J. Mater. Res. **21**, No. 11, 2006.
14. Xiaobo Huang, Liwen Tan, Haeseok Cho and Bethanie J. H. Stadler, J. Appl. Phys. **105**, 07D128 (2009).

Single Spin Dynamics/DMS I
(III-V and Group IV)

Mater. Res. Soc. Symp. Proc. Vol. 1183 © 2009 Materials Research Society   1183-FF05-01

Experimental evidence of the hyperfine interaction between hole and nuclear spins in InAs/Gaas quantum dots

B. Eble,[1] C. Testelin,[1] P. Desfonds,[1] F. Bernardot,[1] A. Balocchi,[2] T. Amand,[2] A. Miard,[3] A. Lemaître,[3] X. Marie,[2] and M. Chamarro[1]

[1] Institut des NanoSciences de Paris, Université P. et M. Curie, CNRS UMR 7588,140 rue de Lourmel, F-75015 Paris, France

[2] Université de Toulouse; INSA, UPS, CNRS; LPCNO, 135 avenue de Rangueil, F-31077 Toulouse, France

[3] Laboratoire de Photonique et Nanostructures, CNRS, Route de Nozay, F-91460 Marcoussis, France

ABSTRACT

The spin dynamics of resident holes in singly p-doped InAs/GaAs quantum dots is studied by pump-probe photo-induced circular dichroism experiments. We show that the hole spin dephasing is controlled by the hyperfine interaction between the hole spin and nuclear spins. We find a characteristic hole spin dephasing time of 12 ns, in close agreement with our calculations based on a dipole-dipole coupling between the hole and the quantum dot nuclei. Finally we demonstrate that a small external magnetic field, typically 10 mT, quenches the hyperfine hole spin dephasing.

INTRODUCTION

Recently, the spin of a single electron localized on a nanometer scale has been proposed as a good candidate for quantum information and quantum computation applications [1-3]. Compared to bulk materials, the strong spatial confinement of carriers in semiconductor quantum dots (QDs) efficiently quenches the main spin relaxation mechanisms while enhancing mechanisms based on carrier exchange and hyperfine interaction. The major obstacle for applications is the fast dephasing induced by the coupling of the electron spin with the random fluctuating nuclear spins [4-7]. For a hole, this Fermi contact coupling is expected to be much weaker because of the p-symmetry of the valence band states [8].

However the corresponding hole spin relaxation or dephasing induced by nuclear spins were not evidenced before in semiconductors because the hole spin dynamics in bulk or quantum well structures is governed by the very rapid spin relaxation mechanisms induced by the strong heavy- and light-hole mixing in the valence bands [9-11]. These spin-orbit related effects are inhibited in QDs due to their fully quantized electronic structure. Recently, a lower limit of tens of nanoseconds was estimated for hole spin relaxation times in CdSe or InAs QDs, but these measurements were limited by the radiative recombination of the photo-created complexes [12,13].

We measured the hole spin dynamics by time-resolved optical orientation experiments in an ensemble of p-doped InAs/GaAs QDs. We demonstrate that the hole spin dephasing is controlled by the hyperfine hole-nuclear spin interaction. We show that this effect relies on the dipole-dipole coupling combined with the mixing of heavy- and light-hole states in QDs. Finally we evidence that a small magnetic field, of the order of 10 mT, quenches the hole spin relaxation.

EXPERIMENTS and DISCUSSION

The studied QD structure was grown by molecular beam epitaxy on a (001) GaAs substrate. The sample consists of 30 planes of self-assembled InAs QDs, separated by 38-nm thick GaAs spacer layers. The QD surface density is about 10^{10} cm^{-2}. The structure was p-modulation doped with a Carbon delta-doping layer (nominal density ~ 2×10^{11} cm^{-2}) located below each QD layer. Photoluminescence spectrum shows a maximum at 1.35 eV.

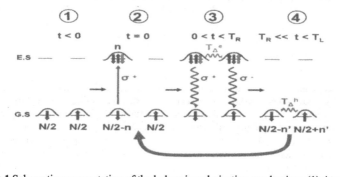

Figure 1 Schematic representation of the hole spin polarization mechanism. (1) At equilibrium, the ground state (G.S.) of a QD is made by a –3/2 or +3/2 hole with equal probabilities, and thus there are as many (N/2) –3/2 and +3/2 holes in the sample. (2) The excited state (E.S.) created from a –3/2 hole by a σ+ pump pulse (solid red arrow) is a positive trion $|\Downarrow \Uparrow \downarrow\rangle$. (3) This photo-generated trion has an electron spin which precesses in the effective nuclear field $\vec{B}_N^e$; this process coherently couples $|\Downarrow \Uparrow \downarrow\rangle$ and $|\Downarrow \Uparrow \uparrow\rangle$ trions, while they recombine (wavy red arrows). T_Δ^e, the electron defasing time, is defined as in ref. [4] an T_R is the lifetime of the trion. (4) Finally, after this recombination, the +3/2 holes are more numerous than the –3/2 ones in the sample; the hole spin can then relax due to the effective nuclear field $\vec{B}_N^h$. Because the hole spin relaxation time is long enough compare to the pump repetition period T_L, the following σ+ pump pulse impinges the sample while the hole polarization is still present (thick horizontal arrow). T_Δ^h is the hole dephasing time and it will defined later in this article.

We measured the photo-induced circular dichroism (PCD) in the QD sample, to probe the resident hole spin polarization. A picosecond Ti:sapphire laser is split into pump and probe beams (the repetition frequency is 76 MHz). The beams propagate along the growth axis z, and are tuned to the energy of the lowest-allowed optical transition of InAs QDs, containing a single resident hole (1.35 eV). The pump beam polarization is σ+/σ– modulated at 42 kHz with a

26

photo-elastic modulator and creates a complex of three particles, called the positive trion (X^+), in its ground state. This transient complex consists of two holes with opposite spins forming a singlet (the photo-generated one and the resident one due to doping), and of a photo-generated electron with its spin pointing down or up depending on the σ+ or σ– circularly polarized excitation, respectively. During the lifetime of these photo-generated electrons, the electron-nuclear hyperfine interaction leads to an efficient coherent coupling of the two electron spin states [4,6]. The spontaneous decay of the trion state by emission of a polarized photon leads to an efficient hole spin cooling, as evidenced recently in cw single QD experiments [14]. This process allows us to spin-polarize the resident holes in the QDs with a pulsed resonant excitation (see Fig. 1).

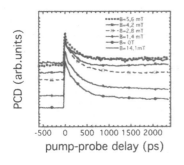
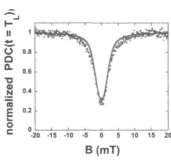

a)

Figure 2 (a) PCD signal as a function of the pump-probe delay, for different values of an external magnetic field $\vec{B}$ applied along the sample growth axis (z direction). The pump and probe beams are tuned to 1.35 eV. The zero-signal level is the same for all displayed curves. T = 2 K. (b) PCD amplitude (full diamonds) at negative pump-probe delay t = -130 ps (*i.e.* t ≈ 13 ns after the previous pump pulse) *versus* the applied longitudinal magnetic field. The solid line is a theoretical fit (see text), in which the maximum hole spin polarization is taken to be equal to 1(parameters $\beta = 0.4$, $T_\Lambda^h = 12$ ns, $g_h = 1.4$, $T_R = 800$ ps, $T_\Lambda^e = 500$ ps and $g_e = 0.4$.).

The probe beam is linearly polarized. After transmission through the sample, it is decomposed into its two circular components, and the difference in their intensities is measured with a balanced optical bridge. Figure 2 (a) shows the temporal behavior of the low-temperature PCD signal obtained when the pump and probe beams are tuned to the trion transition of the p-doped QDs and no magnetic field is applied. The PCD signal has two contributions: (i) the population difference of the spin-polarized heavy-hole ground states $J_z^h = \pm\ 3/2$, and (ii) the population difference of the spin-polarized trion states $J_z^{X^*} = \pm 1/2$. In Fig. 2 (a), we observe a nonzero PCD signal at negative pump-probe delays, indicating that the spin polarization is not fully relaxed within the T_L = 13 ns repetition period of the laser pulses. This long-living component of the PCD signal is unambiguously associated to the net spin polarization of the resident holes, the only species present in the sample after the radiative recombination of trions (of lifetime $T_R \approx 800\ ps$, measured in a time-resolved experiment [15], not shown here).

27

To obtain a direct evidence of the hole-nuclear spin interaction, we have applied an external magnetic field $\vec{B}$ parallel to the growth axis z of the sample. As for an electron [4,5], the hyperfine interaction of a hole in a QD with the surrounding nuclei can be described by a frozen effective nuclear field $\vec{B}_N^h$ acting on the hole spin. This frozen fluctuation approach is justified since the correlation time of the field $\vec{B}_N^h$ (~ 10 μs) is several orders of magnitude longer than the typical hole spin dephasing time, as will be found later. The only difference is the physical origin of this hyperfine interaction. For an electron, it has a dominant Fermi contact character because of the s-symmetry of the wavefunction. However, a hole in the valence band possesses a p-symmetry (the admixture with s-symmetry conduction wavefunction being negligible [16]), the wavefunction at the nuclei vanishes and the contact interaction is thus very weak. This is the reason why the hole-nuclear spin interaction has been neglected so far in QDs. Nonetheless, the nuclei can also interact with carriers through dipole-dipole coupling [17]. In the absence of applied magnetic field, each resident hole spin precesses coherently around the effective nuclear magnetic field $\vec{B}_N^h$, as a consequence of the dipole-dipole hyperfine interaction. The average hole spin polarization in the QD ensemble thus decays with time because of the random distribution of the local nuclear effective fields. In an external magnetic field $\vec{B}$ applied along z, the hole spin dephasing induced by the hyperfine interaction can be strongly reduced if the amplitude of the external field is larger than the dispersion $\Delta_{//}$ of the in-plane fluctuations of the nuclear hyperfine field $\vec{B}_N^h$ (*i.e.* the Zeeman interaction of the hole spin with $\vec{B}$ is stronger than the interaction with nuclei).

Figure 2(a) shows the PCD signal obtained for several values of the applied magnetic field $\vec{B}$. The mean pump power density is ≈ 60 W.cm^{-2}, close to the previously estimated power density of Rabi π–pulses in n-doped InAs QD [18,19]. In Fig. 2(b), the data show the PCD signal measured at delay t = -130 ps, as a function of the applied magnetic field; they show a dip near B=0, with a full width at half maximum (FWHM) $2\Delta = 5 \pm 0.4$ mT. The striking feature is that a small external field has a dramatic impact on the resident hole spin polarization. The observed rise of the measured PCD(t=-130ps ≈ 13ns) signal when B_z increases reflects a significant increase of the hole spin polarization due to a longer hole spin dephasing time. We emphasize that the external magnetic field B_z remains very small (*i.e.* of the order of few mT; the Zeeman splitting of the electron or the hole in this field is about 3 orders of magnitude smaller than $k_B T$ at T = 2 K). Besides, the rapid $\sigma + /\sigma -$ modulation (at 42 kHz) of the pump beam in the experiment prevents any dynamical polarization of the nuclear spins [20]. Therefore, the increase of the average hole spin polarization observed in Fig. 2(b) is interpreted as a consequence of the suppression of the hole spin dephasing induced by the interaction with nuclear spins, the hole spin relaxation time being much longer than the laser repetition period ($T_L \sim 13$ ns). Hence, the PCD signal at negative delays corresponds to an average hole spin polarization which results from the equilibrium between the laser repetition period and the relevant hole spin relaxation time T_1^h (no longer due to the hyperfine interaction with nuclei if B_z is larger than 10 mT).

To confirm this interpretation, we have calculated the hole–nuclear hyperfine interaction through dipole-dipole coupling, and showed that it explains the magnetic field dependence of the PCD signal at 13 ns displayed in Fig. 2(b). Details are given in Ref. [15]. In a QD, confinement and biaxial strains lift the degeneracy between the heavy-hole (hh) and light-hole (lh) states. For pure hh states, the dipole-dipole hyperfine hamiltonian does not couple the $J_z^h = \pm 3/2$ hole spin states. Nonetheless, there are clear experimental evidences of heavy and light hole mixing of the

28

hole wavefunction in self-assembled QDs induced by in-plane anisotropy and strain [21,22,23]. Using the modified hh states [23,24], the dipole-dipole Hamiltonian [17] for a given hole can be written. It is equivalent to an effective nuclear field $\vec{B}_N^h$ acting on the hole magnetic moment. The magnitude and direction of this field are randomly distributed from QD to another QD, and the randomness is described by an anisotropic Gaussian probability distribution of $\vec{B}_N^h$ with $\Delta_{//}$ and $\Delta_\perp$ the quadratic averages of the in-plane and perpendicular-to-the-plane components, respectively. Assuming a simple model with a constant wavefunction inside the QD and equal to zero outside, $\Delta_{//}$ and $\Delta_\perp$ write:

$$g_h \mu_B \Delta_{//} = \alpha g_h \mu_B \Delta_\perp = \frac{\alpha}{1+|\beta|^2} \sqrt{\frac{4\sum_j I^j (I^j + 1)(C_j)^2}{3N_L}} = \frac{\hbar}{T_\Delta^h} \,, (1)$$

where T_Δ^h is the ensemble dephasing time, arising from the random hole precession directions and frequencies in the randomly distributed frozen nuclear field. N_L is the number of nuclei inside a QD, and the summation runs over the nuclei j, with spin I^j, inside a unit cell. The dipole-dipole hyperfine constants C_j can be obtained by using the dipole-dipole matrix elements derived in Ref. [8]. $\alpha = 2|\beta|/\sqrt{3}$ is the anisotropy constant. Typical values of $|\beta| = 0.2 - 0.7$ have been observed in InAs, CdSe or CdTe QDs [21,22,23]. Finally, taking the estimated values $C_{As} = 4.4\ \mu eV$, $C_{In} = 4.0\ \mu eV$, and the value $N_L = 6 \times 10^4$ deduced from time-resolved PL measurements of the spin dynamics of the photo-generated electrons [15], we are able to estimate the hole spin dephasing time T_Δ^h from the above equation; for a typical value of the anisotropy constant observed experimentally [21,22,23] $|\beta| = 0.5$, we find $T_\Delta^h = 14$ ns. That leads to a quadratic average characterized by $2\Delta_{//} = 1.1$ mT (assuming $g_h \approx 1.5$). We note that this value is of the same order of magnitude but smaller than the FWHM of the experimental curve in Fig. 2(b). In fact the width of the experimental curve does not correspond directly to $2\Delta_{//}$: the width increases when the observation time becomes comparable or shorter than the spin dephasing time, as already calculated by Merkulov et al. for the magnetic field dependence of an electron spin polarization [4].

We have modeled our PCD experiment on a p-doped QD sample, and fitted the dependence of the PCD signal at 13-ns delay on an applied longitudinal field $\vec{B}$ (red curve in Fig. 2(b)). For that purpose, as the hole spin polarization is not relaxed within the time interval between two pump pulses, we have considered the coupled spin dynamics of a resident hole and of the corresponding photo-generated electron (i.e. the trion) under periodic excitation by the pump pulses, and sought for a stationary solution where the hole spin polarization just before the arrival of a pump pulse equals the one at delay T_L. After a σ+ pump pulse, supposed to perform an exact π Rabi oscillation (see step 2 in Fig.1), the electron spin evolves in magnetic field $\vec{B} + \vec{B}_N^e$, where $\vec{B}_N^e$ originates from the contact hyperfine electron-nuclear interaction, and the trion recombines (see step 3 in Fig.1). After recombination (during which the hole spin dynamics is neglected $T_R \ll T_\Delta^h$), the hole spin evolves in magnetic field $\vec{B} + \vec{B}_N^h$ (see step 4 in Fig.1). The nuclear fields $\vec{B}_N^h$ and $\vec{B}_N^e$ are linked because they are created by the same nuclei. The hole spin polarization is then evaluated at delay T_L in the stationary regime, and finally averaged over a

Gaussian distribution of $\vec{B}_N^h$ with the preceding $\Delta_{//}$ and $\Delta_{\perp}$ parameters. The fit of the data of Fig. 2(b) is quite good; it yields a reasonable anisotropy parameter $\alpha = 0.5 \pm 0.1$ and the dephasing time $T_\Delta^h = 12 \pm 2 \ ns$, in very good agreement with the above estimated value of 14 ns.

In conclusion, we have demonstrated here that the hole spin dynamics in InAs/GaAs QDs, in the absence of external magnetic field, is governed by the dipole-dipole interaction with random nuclear spins. Its effect on the hole spin relaxation time can be efficiently suppressed by a much smaller external magnetic field, ≈ 10 mT, than the ≈ 200 mT required to screen the hyperfine contact interaction of an electron with nuclear spins in the same QDs [6]. This hole-nuclear spin coupling, which has not been measured before in semiconductors, must be taken into account in future hole-spin-based quantum devices. Finally, studies to control and minimize the heavy-light hole mixing would be interesting in order to reduce or even cancel the hyperfine interaction of holes in QDs.

ACKNOWLEDGMENTS

We acknowledge O. Krebs and B. Urbaszek for fruitful discussions. One of us (B.E.) thanks the C'Nano-IdF for its financial support.

REFERENCES

1. D. Loss, and D. DiVincenzo, *Phys. Rev. A* 57, 120-126 (1998).
2. D. D. Awschalom, , D. Loss, and N. Samarth (eds), *Semiconductor Spintronics and Quantum Computation* (Springer, Berlin, 2002).
3. R. Hanson, et al., *Rev. Mod Phys.* 79, 1217- (2007).
4. I.A. Merkulov, Al L Efros. and M. Rosen, *Phys. Rev. B* 65, 205309,(2002).
5. A.V. Khaetskii, D. Loss and L. Glazman, *Phys.Rev. Lett* 88 186802 (2002).
6. P.-F. Braun et al., *Phys. Rev. Lett.* 94, 116601 (2005).
7. A.C. Johnson et al., *Nature* 435, 925, (2005)
8. E.I. Gryncharova and V.I. Perel *Sov. Phys. Semicond.* 11, 997 (1977).
9.T. Damen et al., *Phys. Rev. Lett.* 67, 3432 (1991).
10. T. Amand et al. *Phys. Rev. B*, 50, 11624 (1994).
11. B. Baylac et al., *Solid. State Comm.* 93, 57 (1995).
12. T. Flissikowski *et al*, *Phys. Rev. B* 68, 161309 (2003).
13. S. Laurent *et al.*, *Phys. Rev. Lett.* 94, 147401 (2005).
14. B.D. Gerardot *et al.*, *Nature* 451, 441 (2008).
15. C. Testelin *et al.* arXiv:0903.3874. Eble *et al.* to be published PRL 102, 146601,(2009)
16. G. Bester, S. Nair and A. Zunger, *Phys. Rev. B* 67, 161306 (2003).
17. A. Abragam, *The Principles of Nuclear Magnetism* (Clarendon, Oxford, 1973), p. 172.
18. A. Greilich *et al.*, *Phys. Rev. Lett.* 96, 227401 (2006)
19. M. Chamarro, F. Bernardot and C. Testelin, *J. Phys. Condens. Matter* 19, 445007 (2007)
20. P. Maletinsky et al., *Phys. Rev. Lett.* 99, 056804 (2007)
21. D.N. Krizhanovskii et al, *Phys. Rev. B* 72, 161312 (2005)
22. A.V. Koudinov et al, *Phys. Rev. B* 70, 241305 (2004)
23. Y. Léger, L. Besombes, L. Maingault and H. Mariette, *Phys. Rev. B* 76, 045331 (2007).

Mater. Res. Soc. Symp. Proc. Vol. 1183 © 2009 Materials Research Society   1183-FF05-02

Electrical Single-Spin Manipulation in Gated Quantum Dots Via Closed Loop Trajectories

M. Fearn and J. H. Jefferson
QinetiQ, St. Andrews Road, Malvern WR14 3PS, England

ABSTRACT

Rotations of a single spin-qubit are demonstrated using only electrical gates. The results of simulations are presented for a typical gated quantum-well structure based on the strong spin-orbit semiconductor InSb, showing typical length and time scales for Hadamard and NOT operations. Approximate analytic expressions are derived for the spin transformations that give excellent agreement with the numerical results and demonstrate the generality of the method.

INTRODUCTION

The controlled manipulation of a single spin is a requirement for implementing universal quantum gates in a quantum information processing device in which the qubits are stored as single spins confined in quantum dots [1]. Impressive progress [2] has been made in recent years in the fabrication and control of electron occupation in quantum dots defined by patterned surface gates in two-dimensional electron gases (2DEG) formed in semiconductor heterostructures and single-electron dot occupancy is now achievable [3]. Recently, the coherent control of a single electron spin in a gated quantum dot via electric fields was demonstrated [4] in which the spin rotations were mediated by spin-orbit coupling. In this experiment the electron confined in the gated quantum dot was displaced periodically around its equilibrium position via the ac field and in an external magnetic field. In the present work, we consider, theoretically and via numerical simulation, a related setup in which the dot is moved in a closed loop geometry via surface gates [5]. This has the potential advantage of being 'all electrical' in that an external magnetic field is not needed to realize spin rotations. These depend only on the size and geometry of the closed loop and are independent of time, provided the adiabatic approximation is valid. Since spin transformations need to be performed within a coherence time, this sets the length scales and timescale for the method to be viable. We derive analytic expressions for general spin rotations in scaled units, which demonstrate the universal nature of the method, and also give realistic estimates for a possible realization based on gated InSb quantum wells.

THEORY & DISCUSSION

We consider a semiconductor 2DEG in the absence of a magnetic field in which the voltage on four surface gates may be varied independently. This is shown schematically in figure 1. Here we assume that the gate voltages may be chosen such that the region between the gates contains a single electron and that the two-dimensional confining potential landscape, $V(x,y,t)$, may be further varied by apply small voltage changes to the gates. An effective Hamiltonian which describes this situation, with time-dependent gate voltages, takes the form [6]:

$$H(t) = \frac{p_x^2 + p_y^2}{2m} + V(x,y,t) - (\beta - \alpha)p_x\sigma_y - (\beta + \alpha)p_y\sigma_x \quad (1)$$

where p_x and p_y are momentum operators in x and y directions and m is the electron effective mass. The last term two terms in equation (1) are the Rashba and Dresselhaus energy resulting from the spin-orbit interaction. The Dresselhaus parameter, β, depends mainly on the quantum well material and may be regarded as approximately constant in this paper, whereas the Rashba parameter α is approximately proportional to the magnitude of an effective electric field in the z-direction inside the quantum dot. This effective electric field may be varied by applying a potential across the quantum well from which the gated quantum dot is formed (not shown in figure 1). In this way the relative magnitudes of $(\beta-\alpha)$ and $(\beta+\alpha)$ in equation (1) may be 'tuned'. The effective 2D hamiltonian (1) may be derived starting with the full 3D hamiltonian and integrating over the lowest confined channel in the z-direction. Explicit

Figure 1 : Schematic of device considered – a quantum dot (QD) in a semiconductor 2DEG defined by surface gates.

numerical estimates of the parameters and the effective 2D potential $V(x,y,t)$ for a gated InSb quantum dot are given later, along with the results of a simulation obtained by explicitly integrating the time-dependent Schrödinger equation. We first give approximate analytic solutions of the time-dependent problem under the assumption that the effect of applying biases on the 4 gates is to translate the quantum-dot confining potential of harmonic oscillator form whilst maintaining its minimum and shape at low energy, ie $V(x,y,0) = V_0(x,y) \to V(x,y,t) = V_0(x - \bar{x}, y - \bar{y})$ where $(\bar{x}, \bar{y})$ is the displacement vector of the dot.

For a one-dimensional system (quantum wire) it has been shown that the Rashba-Dresselhaus terms may be transformed away by a unitary transformation leading to an analytic solution of the Schrödinger equation [7]. In two dimensions they may be removed approximately in a piecewise fashion for linear translations of the dot potential well, provided the translations are performed sufficiently slowly (adiabatic approximation). Consider, for example, a translation of the dot in the x-direction. We first perform a unitary transformation $R_x\left(-2y/l_y\right) = e^{-i\sigma_x y/l_y}$, which gives a spin-rotation $-2y/l_y$ about the x-axis and also a spin-dependent translation of p_y. If we choose $l_y = \dfrac{\hbar}{m(\beta+\alpha)}$, this precisely cancels the fourth term in (1), apart from an unimportant constant, giving the transformed Hamiltonian:

$$\tilde{H}(t) = R_x H R_x^+ = \frac{p_x^2 + p_y^2}{2m} + V_0(x - \bar{x}(t), y) - (\beta-\alpha)p_x\left[\sigma_y\cos(2y/l_y) - \sigma_z\sin(2y/l_y)\right]$$

$$\approx \frac{p_x^2 + p_y^2}{2m} + V_0(x - \bar{x}(t), y) - (\beta-\alpha)p_x\sigma_y \qquad (2)$$

where in the last step we have set $y = \bar{y} = 0$, which is a good approximation under high dot confinement. The last term in equation (3) may now be removed by the further unitary transformation $R_y(-2x/l_x) = e^{-i\sigma_y x/l_x}$, where $l_x = \dfrac{\hbar}{m(\beta-\alpha)}$, giving the final transformed Hamiltonian with Rashba and Dresselhaus terms removed:

$$\hat{H}(t) = R_y \tilde{H} R_y^+ = \frac{p_x^2 + p_y^2}{2m} + V_0(x - \bar{x}(t), y). \qquad (3)$$

32

Note that the order of the two unitary transformations is important since they do not commute and the approximation in (2) would no longer be valid if the rotation about the y-axis were performed first (except for $\bar{x}(t) \sim 0$). In the adiabatic approximation, the ground state of (3) satisfies $\hat{H}(t)\hat{\Psi}_0(t) = E_0(t)\hat{\Psi}_0(t)$

where $\hat{\Psi}_0(t) = \psi_0(x - \bar{x}(t), y)\chi_0$, with $\psi_0(x - \bar{x}(t), y) = \dfrac{1}{2\pi l_0^2} e^{-\frac{[x-\bar{x}(t)]^2+y^2}{2l_0^2}}$ the x-displaced harmonic

oscillator orbital ground state and χ_0 the spin state. The ground state of $H(t)$ is thus, $\Psi_0(t) = R_x^{-1}R_y^{-1}\hat{\Psi}_0(t) = \psi_0(x-\bar{x}(t),y)e^{i\sigma_x y/l_y}e^{i\sigma_y x/l_x}\chi_0$. We see from this equation that the main effect of the adiabatic translation of the dot is to rotate the spin about the y-axis provided $l_x, l_y > l_0$, ie $\Psi_0(t) \approx \psi_0 e^{i\sigma_y \bar{x}/l_x}\chi_0$, a spin rotation of $2\bar{x}/l_x$ around the y-axis.

We now consider the motion of the dot in a closed loop, which we approximate to a rectangle, as shown in figure 2. Between t_1 and t_2 the quantum dot moves adiabatically in the y-direction with spin state $e^{i\sigma_y X/l_x}\chi_0$ at $t=t_1$. For $t>t_1$ we perform unitary transformations as before but this time the rotation about the y-axis is performed first and takes the form $R_y[-2(x-X)/l_x] = e^{-i\sigma_y(x-X)/l_x}$, where $X = \bar{x}(t_1)$. This removes the last term in equation (1) to a good approximation. The ground state of $H(t)$ is then $\Psi_0(t) = \psi_0(x-X, y-\bar{y}(t))e^{i\sigma_y(x-X)/l_x}e^{i\sigma_x y/l_y}e^{i\sigma_y x/l_x}\chi_0$ $(t_2 \geq t \geq t_1)$

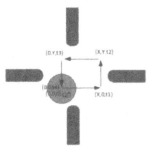

Figure 2 : Schematic of scheme to manipulate a single spin in a quantum dot. The dot is displaced in a closed loop via time-pulsed gate voltages. The space-time co-ordinate (x,y,t) is labeled at each corner of the loop.

The ground state over the two remaining segments may be obtained in a similar way and the complete solution in the high-confinement approximation then takes the form $\Psi_0(t) = \psi_0(x - \bar{x}(t), y - \bar{y}(t))\chi$ where:

$$\chi = e^{i\sigma_x y/l_y}e^{i\sigma_y x/l_x}\chi_0 \approx e^{i\sigma_x \bar{y}/l_y}e^{i\sigma_y \bar{x}/l_x}\chi_0 \qquad (t_1 \geq t \geq 0)$$

$$= e^{i\sigma_y(x-X)/l_x}e^{i\sigma_x y/l_y}e^{i\sigma_y X/l_x}\chi_0 \approx e^{i\sigma_x \bar{y}/l_y}e^{i\sigma_y X/l_x}\chi_0 \qquad (t_2 \geq t \geq t_1)$$

$$= e^{i\sigma_x(y-Y)/l_y}e^{i\sigma_y(x-X)/l_x}e^{i\sigma_x Y/l_y}e^{i\sigma_y X/l_x}\chi_0 \approx e^{i\sigma_x(\bar{x}-X)/l_x}e^{i\sigma_x Y/l_y}e^{i\sigma_y X/l_x}\chi_0 \qquad (t_3 \geq t \geq t_2) \qquad (4)$$

$$= e^{i\sigma_x(y-Y)/l_y}e^{i\sigma_y(x-X)/l_x}e^{i\sigma_x Y/l_y}e^{i\sigma_y X/l_x}\chi_0 \approx e^{i\sigma_x(\bar{y}-Y)/l_y}e^{-i\sigma_y X/l_x}e^{i\sigma_x Y/l_y}e^{i\sigma_y X/l_x}\chi_0 \qquad (t_4 \geq t \geq t_3)$$

In the last step we have made the high-confinement approximation, setting $x = \bar{x}$ and $y = \bar{y}$. After the complete cycle, the spin will thus be transformed to:

$$\chi = e^{i\sigma_x(y-Y)/l_y}e^{i\sigma_y(x-X)/l_x}e^{i\sigma_x Y/l_y}e^{i\sigma_y X/l_x}\chi_0 \approx e^{-i\sigma_x Y/l_y}e^{-i\sigma_y X/l_x}e^{i\sigma_x Y/l_y}e^{i\sigma_y X/l_x}\chi_0 \qquad (5)$$

We see from these equations that the behavior of the model is universal in that it is only the ratios X/l_x and Y/l_y that are important and henceforth we will consider only the Rashba case with $\beta=0$, ie $l_y = -l_x = l = \dfrac{\hbar}{m\alpha}$, the solutions with finite β being simply related by a scaling of length.

Since sequential rotations in (5) are about different axes they do not commute and by appropriate choice of rotation angles we may perform net rotations of the spin. For example, choosing $X/l=\pi/4$, equation (5) becomes,

$$\chi \approx \begin{pmatrix} e^{-iY/l}\cos(Y/l) & ie^{iY/l}\sin(Y/l) \\ ie^{-iY/l}\sin(Y/l) & e^{iY/l}\cos(Y/l) \end{pmatrix}\chi_0 \, ,$$

(6)

a general rotation at constant azimuth ($\phi=\pi/2$) on the Bloch sphere. This gives a Hadamard transformation for $Y/l=\pi/4$ and a NOT operation for $Y/l=\pi/2$. The Hadamard transformation is shown in figure 3a for the full cycle of figure 2 from direct solutions of the time-dependent Schrödinger equation. Numerical solutions were obtained for varying time steps and several grid sizes in real space to test for convergence.

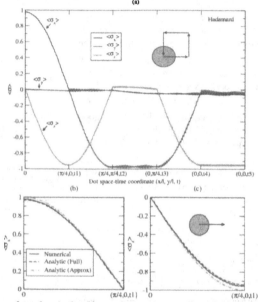

Figure 3 : Hadamard transformation via closed loop trajectory. a) expectation values of spin from solution of the time-dependent Schrödinger equation. Comparison of analytic approaches (see text) and numerics for (b) $<\sigma_x>$ and (c) $<\sigma_z>$ in first segment of trajectory.

In this plot, the time is not shown explicitly but relates to the spatial coordinates by $t=x/v$ or $t=y/v$, where v is the speed of the dot, which for the results shown corresponds to 9.6 nm/ps. To be specific, we may relate these parameters to plausible values for InSb in which we include only the Rashba term, choosing $\alpha=7.6\times10^4$m/s giving $l=109$nm for an effective mass of $0.014m_e$. The oscillator energy of the harmonic confining potential is $\hbar\omega_0=15$meV, giving a confinement length of the quantum dot of $l_0=\sqrt{\hbar/m\omega_0}=19$nm. Hence $l/l_0=5.7$, showing reasonably high confinement. This is supported by figure 3b, which is a magnified plot of $<\sigma_x>$ along the first segment of the trajectory, and similarly for $<\sigma_z>$ in figure 3b. Also shown in these plots is the infinite confinement limit (green curves) given directly from the approximation in equation (4). These results may be determined analytically and we get:

$$\langle\sigma_x\rangle = \cos\frac{\pi}{2}\frac{x}{X}, \quad \langle\sigma_y\rangle = 0, \quad \langle\sigma_z\rangle = -\sin\frac{\pi}{2}\frac{x}{X}$$

and show that the infinite confinement limit reproduces the main spin rotation effects but that the errors are not insignificant. Results using the more accurate expressions for the spinor in equation (4) are also plotted (red curves) and show excellent agreement with the numerical simulation. Furthermore, numerical simulations (not shown here) which relax the condition that the shape and depth of the well remains constant during the closed loop trajectory show that the results are only weakly perturbed. Since $l_0\omega_0$=433nm/ps is significantly greater than v, the adiabatic approximation should be reasonable and this is supported by the numerical results in figure 3 which, nevertheless, shows weak oscillations at frequency $\sim\frac{1}{2}\omega_0$, that may be understood directly from lowest-order time-dependent perturbation theory.

It is clear from the behavior of the spin components in figure 3, and the approximate expressions in equation 4, that the Hadamard transformation results from first rotating the spin through a right angle about the y-axis so it now point in the $-z$ direction. Subsequent motion along y then rotates the spin a further right angle about x thus aligning it along $-y$. Since the next segment rotates the spin about the y-axis, it cannot change its direction (only its overall phase) and in the final segment it then rotates about x to finally point in the direction of $-z$. In figure 4, for comparison, we show numerical results for the NOT operation for which we choose $Y/l=\pi/2$ (cf equation 6), These again gives excellent agreement with results from equation 4.

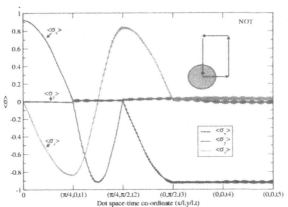

Figure 4 : NOT operation via closed loop trajectory.

Note that in this case, the spin is rotated to point in the $-x$ direction after the third segment and hence cannot change its direction in the last segment. Intermediate rotations may be achieved by varying Y in the closed loop as indicated in equation (6), and more generally by independently varying X.

35

CONCLUSIONS

In this paper we have demonstrated the feasibility of spin rotations by closed-loop trajectories, without external magnetic field, to perform spin rotations that depend on the topology and absolute size of the closed-loop trajectories, independent of time provided the adiabatic approximation is satisfied. Results from direct numerical simulation of the time-dependent Schrödinger equation have been justified analytically with approximate expressions that give excellent agreement with the numerical results. Although the behavior is generic, specific dimensions relevant to material parameters for InSb have been presented and shown to be within current fabrication capabilities for gated quantum well devices.

ACKNOWLEDGMENTS

The authors acknowledge financial support from the UK Ministry of Defence.

REFERENCES

1. D. Loss and D.P. DiVincenzo, Phys. Rev. A, **57**, 120 (1998).
2. J.M. Elzerman, R. Hanson, J.S. Greidanus, L.H. Willems van Bevern, S. De Franceschi, L.M.K. Vandersypen, S. Tarucha, L.P. Kouwenhoven, Phys. Rev. B, **67**, 161308(R) (2003).
3. R.Hanson, L.P. Kouwenhoven, J.R. Petta, S. Tarucha, L.M.K. Vandersypen, Rev Mod Phys, **79**, 1217 (2007).
4. K.C. Nowack, F.H.L. Koppens, Yu. V. Nazarov, L.M.K. Vandersypen, Science, **318**, 1430 (2007).
5. S. Bednarek and B. Szafran, Phys. Rev. Lett., **101**, 216805 (2008).
6. M. Pletyukhov, A. Shnirman, Phys. Rev. B, **79**, 033303 (2009).
7. C. Flindt, A.S. Sorensen, K. Flensberg, Phys. Rev. Lett., **97**, 240501 (2006).

Mater. Res. Soc. Symp. Proc. Vol. 1183 © 2009 Materials Research Society   1183-FF05-09

Four stable magnetization states formed in a single layer of GaMnAs ferromagnetic film

Sanghoon Lee[1], X. Liu[2], and J. K. Furdyna[2]
[1]Department of Physics, Korea University, Seoul, 136-701, Korea
[2]Department of Physics, University of Notre Dame, Notre Dame, 46556, USA

ABSTRACT

GaMnAs ferromagnetic semiconductor films under compressive strain are characterized by strong biaxial in-plane anisotropy, which generates four stable magnetization directions at zero magnetic field. This feature results in double switching behavior during magnetization reversal process measured by planar Hall resistance (PHR). Minor scans of the PHR exhibit staggered asymmetric loops due to the formation of stable muti-domain structures. We show that the resulting four stable PHR states can serve as quaternary logic states for a spin memory device.

INTRODUCTION

The phenomenon of giant magnetoresistance (GMR) in magnetic multilayers made it possible to achieve diverse spintronic device applications, [1] the best known example being the magnetic random access memory (MRAM) based on ferromagnetic metal multilayer system. [2] Although this binary memory device is already successfully fabricated and commercialized, it is desirable to further improve such devices either by using simpler structures, or by enhancing the storage capacity via the concept of multinary states. The multinary-state concept was developed based on magnetic tri-layer structures consisting of two magnetic layers with a separating nonmagnetic barrier. [3] Four different magnetization states can be realized in such a trilayer by independently controlling the magnetization of each magnetic layer by an appropriate external magnetic field, that leads to parallel or anti-parallel magnetic configuration in the trilayer. This approach, however, involves rather complicated trilayer spin valve structures, and also requires multiple steps in the application of magnetic field for setting the desired configuration of layer magnetizations. Furthermore, since the reading process can alter the existing configuration of magnetization orientations, it is also necessary to restore the state of magnetization after the sensing process by another application of magnetic field. Such complicated writing and reading processes, together with the complexity of the trilayer structure, are major drawbacks of the trilayer ferromagnetic system for utilizing the four spin configurations in practical devices. In this study we present another approach for realizing a stable quaternary magnetization state based on a much simpler *single layer* GaMnAs ferromagnetic system.

EXPERIMENT

Epilayers of the ferromagnetic semiconductor GaMnAs were prepared by molecular beam epitaxy (MBE) in a Riber 32 R&D MBE machine equipped with elemental sources Ga, Mn, and As. A GaAs (001) substrate was transferred to the MBE preparation chamber and deoxidized by heating to $620°\,C$ for several minutes. A thick GaAs buffer layer was first grown at $600°\,C$, and the substrate was then cooled to $220°C$ for deposition of a 30 nm low-temperature (LT) GaAs buffer. On top of LT-GaAs, GaMnAs layer was grown with desired Mn concentration and thickness by controlling Mn cell temperature and growth time, respectively. A 100 nm $Ga_{1-x}Mn_xAs$ film with $x \approx 0.038$ was studied in this work. For Hall experiments, a Hall bar was patterned from the GaMnAs epilayer using photolithography and chemical wet etching. The Hall bar is in the form of a rectangle 0.4 mm wide and 2.5 mm long, with the long dimension along the [110] direction and with six terminals for metal contacts, as shown in Fig. 1(a). Hall measurements were performed using a sample holder which permitted a magnetic field to be applied in the plane of the sample at an arbitrary azimuthal angle. All transport measurements were performed at 4 K.

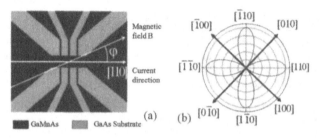

Figure 1 **(a)** Schematic view of Hall bar patterned from the GaMnAs film. The long dimension is along the [110] direction, as indicated in the figure. The dark- and light-colored regions correspond to GaMnAs film and GaAs substrate, respectively. **(b)** Polar plot of magnetic energy, indicating easy axes.

RESULTS AND DISCUSSION

The magnetic behavior of GaMnAs epilayers [4-8] can be understood in terms of magnetic energy of a single in-plane magnetic domain, given by [9,10]

$$E = -K_c \cos^2(2\varphi)/4 + K_u \sin^2 \varphi - MH \cos(\theta - \varphi) \qquad (1),$$

where K_c and K_u are the cubic and uniaxial magnetocrystalline anisotropy constants; H is the external magnetic field; M is the magnetization; and φ and θ are the angles of H and M relative to the [110] direction (i.e., the direction of the current in this study), respectively. While the uniaxial anisotropy term provides only two energy minima, the cubic term gives four minima due to the 2φ dependence of Eq. (1). Thus the magnetic energy minima, which determine the easy axes of magnetization of the ferromagnetic system at zero magnetic field, depend *on the relative strengths* of the two anisotropy energy terms K_c and K_u in Eq. (1). Since cubic anisotropy is much stronger in GaMnAs than uniaxial anisotropy (i.e., $K_c > K_u$) at low temperatures, we will, as a first approximation, ignore the contribution of the uniaxial anisotropy. The plot of magnetic energy with dominant cubic anisotropy in the absence of H is shown in Fig 1(b). The magnetic energy clearly shows four stable states of M: along [100], [010], [$\bar{1}$00], and [$0\bar{1}$0] crystallographic directions, i.e., along the four in-plane magnetic easy axes determined by

38

cubic anisotropy. This simple model provides a good description of the double hysteresis shown in Fig. 2(a) in terms of a two-step switching of M between these easy axes.

The double switching behavior in the main hysteresis loop of the planar Hall resistance (PHR) depicted in Fig. 2(a) is similar to the behavior of magnetoresistance (MR) observed in a spin valve system based on a ferromagnetic trilayer structure. Since the value of PHR is determined by the angle between the magnetization and the current [9], four magnetization states can be distinctly identified in the spectrum, as indicated by the directions of the arrows in Fig. 2. Figures 2(b) and 2(c) illustrate minor loops of the PHR hysteresis spectrum observed in the mode in which only the 90 degree rotation of magnetization is experienced. This is achieved by limiting the field scan to the region between the first (H_{c1}) and the second (H_{c2}) switching fields indicated in Fig. 2(a). In the minor loop shown in Fig. 2(b) the system was initially saturated in the "negative" direction by a field of -1000 G, and the field scan was carried out in the range of ±500 G; while the analogous loop shown in Fig. 2(c) was obtained for an initial saturation in the "plus" direction by +1000 G. As one can see from the two minor loops of the PHR, there are *four* stable states, corresponding to four magnetization orientations at zero magnetic field. These four magnetizations provide a quaternary storage mechanism using only a single ferromagnetic layer, as discussed below.

The four logic states '1', '2', '3', and '4' can be defined as the four magnetizations of the GaMnAs layer, along [$\bar{1}$00], [0$\bar{1}$0], [100], and [010] directions, respectively. These magnetizations are represented by thick arrows in Fig. 2. Each of the four states can be easily set by using appropriate magnetic field strengths, defined as $H_2 > H_{c2}$ and $H_{c2} > H_1 > H_{c1}$. For example, applying magnetic field H_2 or $-H_2$ to a device results in programmed states '1' or '3', respectively. The remaining two states can be written using two-step sequences. That is, for programming state '2', one first applies the field H_2 (which sets state '1') followed by $-H_1$. Similarly, state '4' can be written by sequential application of $-H_2$ and H_1.

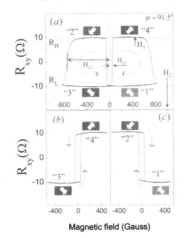

Figure 2. Full hysteresis loop of the planar Hall resistance R_{xy} is shown in panel (a). Minor hysteresis loops obtained with negative and positive initial saturation fields are shown in panels (b) and (c), respectively.

Since the high and the low PHR values defined as R_H and R_L in Fig. 2 have positive and negative values, respectively, the states corresponding to R_H and R_L can be identified by simply measuring the sign of PHR of the device. However, there are two logic states assigned to each value of R_H and R_L, and another identifying process is required to determine the exact state. Each of the four magnetic states can be separately identified in two comparison steps given below:

Step 1: B = 0; measure the sign of PHR;

Step 2: B = H_1 or B = -H_1: measure the sign of PHR.

Details of the sensing processes for the four states are given in Table I.

Table I. Illustration of sensing process of four states

State	Step	B	Result	Restore bit after sensing?
"1"	1	0	PHR < 0	Not needed
	2	H_1	PHR < 0	
"2"	1	0	PHR > 0	Apply B = -H_1
	2	H_1	PHR < 0	
"3"	1	0	PHR < 0	Apply B = -H_1
	2	H_1	PHR > 0	
'4'	1	0	PHR > 0	Not needed
	2	H_1	PHR > 0	

The writing and reading process just described demonstrates the four-state spin memory functionality of a GaMnAs ferromagnetic film with a strong in-plane bi-axial anisotropy. This approach of using PHR in GaMnAs has significant advantages over the tri-layer spin valve structure, since the four states can be realized in *a single layer* of GaMnAs, which significantly simplifies the structure of the four-state spin-memory device.

However, even though a four-state memory device with advantages of a simpler structure can be achieved by using PHR in GaMnAs, this alone offers no improvement over the trilayer spin valve system for the writing and reading process. The four spin orientations determined by magnetic anisotropy in a GaMnAs film have a one-to-one correspondence with the four anti-parallel and parallel spin configurations that can be realized in a trilayer spin valve structure, so that the process of writing and reading is identical in both systems. The main reason for the complicated reading and restoring-after-reading processes is the degeneracy of states. For example, there are two spin combinations for each case of parallel ($\leftleftarrows$ and $\rightrightarrows$) and anti-parallel ($\rightleftarrows$ and $\leftrightarrows$) configurations in the trilayer spin valve system. Similarly, the PHR values for the two collinear spin orientations are the same in the case of a GaMnAs film. Such degeneracy of two states requires complicated processes of reading and restoring, as shown in Table I. However, this complication of reading and restoring can be circumvented by establishing four different PHR values in the GaMnAs film.

The four values of PHR in GaMnAs can be established in the minor loop that was obtained by adjusting the scan range of magnetic field, in which the turning point is within the second magnetization switching region. Figure 3(a) shows such a minor loop spectrum, in which four distinct PHR states are clearly obtained by means of a staggered asymmetric hysteresis. The intermediate states designated as R_{I1} and R_{I2} are multi-domain states, whose stability was recently discovered in the magnetization reversal process of a GaMnAs layer. [11] Based on such asymmetric PHR loops, one can realize four different well-defined states of PHR at zero magnetic field. Since the degeneracy is removed in such an asymmetric loop, the reading procedure becomes very simple, because the state can now be identified just by reading the value of PHR. Thus the realization of asymmetric PHR loops greatly simplifies the complicated process of reading and restoring, which was a major drawback of the four-state memory function both in trilayer spin valve structures and in single-layer GaMnAs.

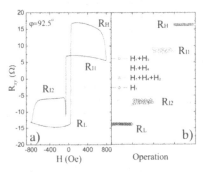

Figure 3. (a) A symmetric PHR loops obtained with φ = 92.5°, showing the four PHR states. (b) Demonstration of writing process given in Table II. The consistent value obtained by each writing process in 30 repetitions indicates that the information can be correctly stored in these four PHR states.

Table II. Sequence of applied magnetic field pulses to write four PHR states.

Logic state	PHR state	Sequence of applying magnetic field H	Magnetic field
"1"	R_H	$H_1 \rightarrow H_3$	$H_1 = -1500G$
"2"	R_{I1}	$H_1 \rightarrow H_4$	$H_2 = -400G$
"3"	R_{I2}	$H_1 \rightarrow H_4 \rightarrow H_2$	$H_3 = +400G$
"4"	R_L	H_1	$H_4 = +800G$

We have further demonstrated the reproducibility of the writing process of the four distinct states shown in Fig. 3(a) as follows. To write the information for each state, we used four different values of magnetic field pulses H_1, H_2, H_3, and H_4 listed in Table II. The application of H_1 or $H_1 \rightarrow H_3$ to a device results in a programmed R_L or R_H state, respectively. To set the intermediate states R_{I1} and R_{I2}, sequences of three and two magnetic field pulses are required, as listed in Table II. We have tested the writing process given in Table II for each of the four states independently, by repeating each writing sequence 30 times. The results are plotted in Fig. 3(b), where all repetitions result in the same value of PHR. This demonstrates that the writing process given in Table II works properly in the GaMnAs device, and that the four PHR states can be used as four spin states for storing information.

CONCLUSIONS

We have demonstrated a four-state memory functionality of the PHR in GaMnAs films with a strong biaxial in-plane anisotropy, that results in four stable independent states of magnetization aligned along the easy axes of the film. This indicates that spin memory device can be fabricated using a *single layer* of GaMnAs – a much *simpler structure* than the current spin-valve-based trilayer structures. Such four distinct values of PHR in GaMnAs are obtained due to the formation of stable multi-domain structures in the transition region of magnetization. These results further simplify the complicated reading-and-restoring processes of the four-state spin memory function, and pave the way toward the realization of elegant high-capacity spin memory devices based on the ability to establish four PHR states in ferromagnetic semiconductor films.

ACKNOWLEDGMENTS

This work was supported by the Korea Science and Engineering Foundation (KOSEF) Grant funded by the Korean Government (MEST) (No. R01-2008-000-10057-0); by the Seoul R&DB Program; and by the National Science Foundations Grant DMR06-03762.

REFERENCES

1. J. M. Daughton, A. V. Pohm, R. T. Fayfield and C. H. Smith, J. Phys. D: Appl. Phys. 32 R169 (1999).
2. Jian-Gang Zhu, Youfeng Zheng, and Gary A. Prinz, J. Appl. Phys. 87, 6668 (2000).
3. JhigangWang and Yoshihisa Nakamura, J. Appl. Phys. 79, 6639 (1996).
4. U. Welp, V. K. Vlasko-Vlasov, X. Liu, J. K. Furdyna, and T. Wojtowicz, Phys. Rev. Lett. 90, 167206 (2003).
5. H. X. Tang, R. K. Kawakami, D. D. Awschalom, and M. L. Roukes, Phys. Rev. Lett. 90, 107201 (2003).
6. A. W. Holleitner, H. Knotz, R. C. Myers, A. C. Gossard, and D. D. Awschalom, Appl. Phys. Lett. 85, 5622 (2004).
7. K. Y. Wang, K. W. Edmonds, R. P. Campion, L. X. Zhao, C. T. Foxon, and B. L. Gallagher, Phys. Rev. B 72, 085201 (2005).
8. H. X. Tang, S. Masmanidis, R. K. Kawakami, D. D. Awschalom, and M. L. Roukes, Nature 431, 52 (2004).
9. K. Okamoto, J. Magn. Magn. Mater. 35, 353 (1983).
10. L. V. Titova, M. Kutrowski, X. Liu, R. Chakarvorty, W. L. Lim, T. Wojtowicz, J. K. Furdyna, and M. Dobrowolska., Phys. Rev. B 72, 165205 (2005).
11. D. Y. Shin, D.Y. Shin, S. J. Chung, Sanghoon Lee, X. Liu, J. K. Furdyna, Phys. Rev. Lett. 98, 047201 (2007)
12. S.J. Chung, D.Y. Shin, Hyunji Son, Sanghoon Lee, X. Liu and J.K. Furdyna, Sol. Sta. Comm. 143, 232 (2007).

Poster Session:
Spintronics Materials

Mater. Res. Soc. Symp. Proc. Vol. 1183 © 2009 Materials Research Society   1183-FF06-01

Ferromagnetism and Near-Infrared Luminescence in Neodymium and Erbium Doped Gallium Nitride Via Diffusion

M. Oliver Luen[1], N. Nepal[1], P. Frajtag[2], J. M. Zavada[1], EiEi Brown[3], U. Hommerich[3], S. M. Bedair[1] and N. A. El-Masry[2]

[1]Electrical and Computer Engineering, North Carolina State University, Raleigh, NC 27695 USA
[2]Materials Science and Engineering, North Carolina State University, Raleigh, NC 27695 USA
[3]Department of Physics, Hampton University, Hampton, VA 23668 USA

ABSTRACT

In this study, we report on the diffusion of neodymium (Nd) and erbium (Er) into n-type and undoped GaN and subsequent measurements of the room-temperature (RT) magnetic and optical properties. The diffusion profile has been measured via secondary ion mass spectroscopy (SIMS) with rare-earth (RE) concentration yields of up to $1 \times 10^{18}/cm^3$. The ferromagnetic properties were measured using an alternating gradient magnetometer (AGM) giving a saturation magnetization (Ms) of up to $3.17 emu/cm^3$ for the RE-diffused layer. The photoluminescence (PL) emission of the Nd-diffused and Er-diffused GaN is observable in the near-infrared (NIR) and infrared (IR) regions of the spectrum, respectively. The Nd-diffused GaN samples show NIR emission at 1064nm and 1350nm, while Er-diffused GaN samples have IR emission at 1546nm. This appears to be the first successful result of Nd diffusion doping into GaN crystals, and the first demonstration of above RT ferromagnetism involving GaN diffused with Nd. Details of our ferromagnetic and optical emission studies, related to the RE diffusion into GaN, are presented.

INTRODUCTION

Rare earth (RE) doped GaN is attracting wide-spread attention both as a diluted magnetic semiconductor (DMS) material and for optical devices useful in telecommunications and multi-color semiconductor display technology. Transition-metal-doped GaN has been reported as a DMS with observable ferromagnetism (FM) in Mn-doped GaN with the Curie temperature above room temperature (RT) [1,2]. Here, we report the above RT ferromagnetic properties of Nd-doped and Er-doped GaN via thermal diffusion of the RE atoms into MOCVD-grown GaN. Wide band gap semiconductors are suitable hosts for the near-infrared (NIR) (1364nm) emission from Nd^{3+} ions corresponding to the low-dispersion wavelength, and infrared (IR) (1546nm) emission from Er^{3+} ions corresponding to the low-loss wavelength for silica-based optical fiber communications, respectively. We report also on the NIR emission seen in our Nd-diffused GaN samples, as well as the IR emission from our Er-diffused GaN samples.

EXPERIMENT

Undoped and Si-doped GaN templates were prepared by MOCVD on (0001) sapphire substrates. Ammonia and silane served as active N and Si sources, respectively. Undoped templates served as baseline experiments for RE diffusion into GaN. The chemically inert 4f electrons of the RE atoms do not enhance conductivity, so thermal diffusion of RE atoms into n-type templates was performed to improve epilayer conductivity for device applications. All GaN:Si films were grown on top of bulk GaN under nitrogen-rich conditions. Si-doping was varied across the templates from low to high with silane flows of 0, 1.5, 3 and 6sccm in order to investigate the effects of template quality on the resultant magnetic and optical characteristics of the RE-diffused GaN:Si. The templates were loaded into a vacuum chamber ($\sim 2 \times 10^{-9}$ torr) containing elemental Nd and Er targets. A thick ~ 8000Å RE metal film was deposited on the GaN templates by pulsed laser ablation (PLA) using a KrF excimer laser at 248nm. The GaN:RE was annealed in-situ for 2-15 hours at 800°C. After wet etching to remove the deposited RE-film and its oxide, RT magnetic measurements were performed using an alternating gradient magnetometer (AGM.) Magnetic measurements were carried out at field strengths of -5kOe to 5kOe for both the in-plane (parallel to sample surface) and out-of-plane (perpendicular to sample surface) orientations. All samples demonstrated FM over the range of Si-doping. The results presented here only include a representative subset of all the samples we produced. PL measurements at NIR and IR wavelengths were achieved via a Nd:YAG laser and Ti-Sapphire laser for GaNdN:Si and GaErN:Si films, respectively.

RESULTS

The profile for Nd concentration in GaN:Si after thermal diffusion was analyzed by secondary ion mass spectroscopy (SIMS.) Figure 1 shows the Nd concentration profile in the diffused samples after 15 hour annealing at 800°C. The Si-doping for these samples was at silane flows of 1.5sccm and 3sccm. The diffusion depth was estimated to be ~ 16nm for the template grown with 3sccm silane, and ~ 32nm for the template grown with 1.5sccm silane. The collected counts/second (cts/sec) for diffused Nd was dependent on Si-doping with a peak of ~ 225cts/sec of Nd for 3sccm silane, and a peak of ~ 2220cts/sec of Nd for 1.5sccm silane. The Nd concentration was estimated from an implanted Nd in Si reference profile to be $\sim 2 \times 10^{17}$/cm³ and 1×10^{18}/cm³ for the 3sccm and 1.5sccm silane templates, respectively.

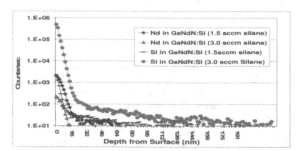

Figure 1. Concentration profile for Nd-diffused samples after 15 hour annealing for 1.5 and 3sccm silane flow. Nd and Si have similar concentrations for 1.5sccm silane flow. Doubling silane flow (3sscm) reduces Nd concentration by 1 order of magnitude.

These low concentrations have been found to be sufficient for ferromagnetic properties from RE ions[3]. No SIMS measurements were carried out for the Er-diffused samples since RE atoms are assumed to have similar diffusion behavior in GaN based on their similar atomic size. Ting *et al.* performed SIMS measurements on GaErN via thermal diffusion in a furnace at 800°C under a flowing nitrogen atmosphere for 7.5-119 hours, and calculated the diffusion coefficient (at 800°C) of Er in GaN through an Arrhenius expression to be $D = 2 \pm 1.1 \times 10^{-17}$ cm²/s[4,5]. We assumed a smaller diffusion coefficient for Nd in GaN due to Nd's larger atomic radius. Pearton *et al.* suggested that impurities like oxygen migrate along threading dislocations in GaN via an interstitial-assisted diffusion mechanism[6]. This is a possible explanation for the migration of RE atoms in GaN films. Ting *et al.*, also claim that the diffusion mechanism for RE metals in GaN is an interstitial-assisted process[4].

Based on the Er concentration profiles by Ting *et al.*, we estimated that in our samples Nd-diffusion depth would be no more than 50nm for our maximum annealing time of 15 hours. SIMS results for our Nd-diffused GaN:Si (Figure 1) showed a dependency of Nd diffusion depth on Si-doping with higher Si-doping leading to lower Nd diffusion and incorporation. Reducing silane flow from 3sccm to 1.5sccm increased Nd concentration by one order of magnitude. The estimated Nd^{3+} ion concentration of $\sim 1 \times 10^{18}/cm^3$ (for 1.5sccm silane) gives a Nd incorporation of <0.01% in the GaN:Si lattice.

It is assumed that the chemically inert 4f electrons of the RE elements produce localized magnetic moments coupled via an indirect exchange interaction mediated by conduction band electrons. This is known as the Ruderman-Kittel-Kasuya-Yosida (RKKY) interaction[7,8]. The unpaired electrons of the highly localized 4f-states give the REs their magnetic properties, resulting in a magnetic contribution of $3\mu_B$ per Nd^{3+} ion. All REs, except Ce in CeN, are trivalent in the nitride system[9]. The RE concentration needed to induce a ferromagnetic response in doped GaN is lower significantly than that required for transition metal doped GaN. Doping of GaN with Mn and Cr requires concentrations on the order of $5 \times 10^{19}/cm^3 - 1 \times 10^{20}/cm^3$ to yield FM while concentrations of $\sim 10^{16} - 10^{18}/cm^3$ are sufficient for a RE species like Gd. Lower doping concentrations lead to less compromise in material quality and allow the host material lattice to be co-doped with conventional shallow dopants to control conductivity[3].

All samples showed ferromagnetic behavior in both the in-plane and out-of-plane directions, with the in-plane saturation magnetization (Ms) significantly stronger than the out-of-plane measurements. This suggests a ferromagnetic canted easy axis in most cases. Figure 2 shows hysteresis curves (Ms (emu) vs. applied field (Oe)) for Nd diffused in undoped and high Si-doped GaN:Si and the inset shows the effect of silane co-doping on the Ms (Ms (emu) vs. silane flow (sccm))of Nd-diffused samples. Each Nd-doped GaN:Si epilayer displayed RT hysteretic behavior consistent with ferromagnetic ordering. The coercivity of the samples was ~200Oe – 300Oe.

Figure 2's inset shows a decrease in Ms for increased silane flow during GaN:Si template growth. Increased Si-doping decreased the Ms due to competition between the Si and RE for Ga sites. This effect was seen in the SIMS data where high Si concentration corresponds to low RE concentration for the same annealing times. The competition arises because V_{Ga} formation energy decreases with increased Si-doping which moves the Fermi level towards the CB[10]. Initially, the slow diffusing Si atoms occupy available V_{Ga} sites yielding n-type GaN, but also create conditions in the host lattice which favor V_{Ga} formation. The added V_{Ga} sites offer the subsequently diffused RE atoms a substitutional lattice site bringing about ferromagnetic properties. At much higher Si-doping, there is too much substitutional Si present and the lowered formation energy for V_{Ga} is less significant, resulting in fewer substitutional sites for the diffused RE atoms. Since Si is introduced to the lattice first (via MOCVD) excess Si can occupy free Ga sites first, during the annealing stage, and rob RE atoms of potential V_{Ga} sites, lowering Ms.

Like Nd, the occupancy of the highly localized 4f-states gives Er a magnetic contribution of $3\mu_B$ per Er^{3+} ion when substituting for Ga in the GaN lattice. Chemical similarities between the two RE species infer similar behavior under comparable experimental conditions. The Er-diffused samples investigated the effect of annealing time (3, 4, 9, 12 and 15hrs) on Ms for constant Si-doping (3sccm silane) at 800°C.

47

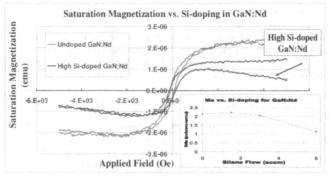

Figure 2. RT hysteresis curves for Nd-diffused GaN:Si with silane flows of 0sccm and 6sccm. Inset: Increased Si-doping decreases the substitutional Nd level yielding lower Ms.

In Figure 3, we found that Ms increased with annealing time up to 9 hours and then appeared to saturate between 9 and 15 hour annealing times. One possible explanation of this result would be a localized high Er concentration (due to shallow diffusion) causing anti-ferromagnetic/ ferromagnetic competition between additional, nearby RE ions. This competition is proposed to occur between substitutional and interstitial RE ions[11].

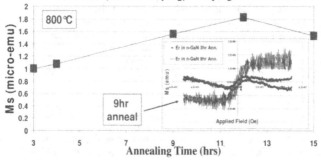

Figure 3. Ms vs. annealing time for Er diffused into GaN:Si templates with similar Si-doping. Inset: RT hysteresis curves for Er-diffused GaN:Si (3, 9hrs).

GaN is a suitable host for the RE atoms with significant ionic bonding which aids the 4f-4f transition probabilities[12,13]. Figure 4 shows the NIR emission spectra measured via an InGaAs detector of the Nd diffused into GaN:Si samples containing varying Si-doping. Excitation of Nd was at 578nm via a Nd:YAG (MOPO) laser. PL was taken using a 0.3m spectrometer with 1μm grating. The InGaAs detector used a RG1000 filter through a 3mm slit. The scan range was from 950nm – 1550nm at 100nm/min. The inset figure shows the emission intensity's dependence on Si-doping (emission intensity (a. u.) vs. silane flow (sccm)). Emission intensity of the Nd-diffused samples is low since we are unable to control precisely the RE concentration using diffusion doping. With RE doping of GaN via MOCVD more efficient luminescence emission is achieved due to better control of the RE concentration throughout the film, which determines emission intensity[14-16]. The PL intensity of the diffused samples shows two peaks in the NIR region. The largest peak corresponds to the $4F_{3/2} - 4I_{11/2}$ transition of Nd^{3+} at ~1064nm, and the other peak corresponds to the $4F_{3/2} - 4I_{13/2}$ transition at ~1350nm. The weak NIR emission intensity of the 1350nm peak might be due to the relatively low concentration of Nd in the diffused GaNdN:Si layer. It has been reported that the optimum concentration for NIR emission from RE-doped GaN is about 0.1-1%[17]. SIMS measurements show our Nd^{3+} concentration to be ~$1 \times 10^{18}/cm^3$ to a depth <50nm this puts our Nd incorporation at ~0.01%, which is below the lower bound of what is considered optimal for good RE emission. At higher RE concentrations, a decrease in emission intensity might be due to decreased transmittance and possibly quenching associated with greater proximity of the RE atoms[18].

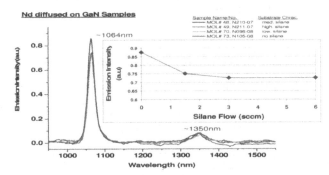

Figure 4. Emission intensity of Nd-diffused GaN:Si. The ~1064nm and ~1350nm peak intensity varies slightly with Si-doping. Inset shows emission intensity at ~1064nm vs. Si-doping (0 - 6sccm).

Our results show repeatable emission at 1064nm and 1350nm from Nd in GaNdN:Si. The peak intensity is dependent on the Si-doping conditions and its sharpness confirms that the emission is intra-atomic and does not depend on the position of the Nd atom within the host lattice. Recall that increased Si-doping decreased Ms due to competition between the Si and Nd for Ga sites, and SIMS showed decreased Nd for increased Si concentration in GaN. This is consistent with our decreased Nd emission intensity as Si-doping is increased.

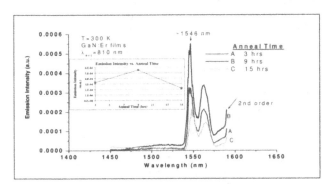

Figure 5. Emission intensity of Er-diffused GaN:Si. The ~1546nm main peak intensity depends on annealing time. Satellite peaks occur from other intra-ff transitions. Inset: emission intensity decreases with longer annealing times

Figure 5 shows the IR emission spectra for Er-diffused GaN:Si and the inset shows emission intensity (a. u.) vs. annealing time (hrs). The excitation source is a 810nm Ti-Sapphire laser. PL

49

was taken using a 0.3m spectrometer with 2μm grating. The InGaAs detector used a RG1000 filter through a 3mm slit. The scan range was from 1450nm – 1590nm at 20nm/min. The main peak corresponds to the $4I_{13/2} - 4I_{15/2}$ transition of Er^{3+} at ~1550nm. Based on the Nd-diffused samples, we assume low Er incorporation in the GaN:Si lattice. The silane conditions were kept constant at 3sccm and the annealing time included 3, 4, 9, 12 and 15 hours. In Figure 4 we saw that Si-doping does affect emission intensity and here we see that longer annealing times decrease the emission intensity of the samples. Emission intensity increases with annealing time up to 9 hours. This confirms the emission intensity's dependence on RE concentration in the host lattice[16]. Beyond 9 hours, we see a decrease in intensity. It is unclear what mechanism causes the loss in emission intensity seen with longer annealing times.

CONCLUSION

In summary, we have shown that n-type GaN:Si grown by MOCVD and ex-situ doped with Nd and Er via diffusion exhibits RT ferromagnetism and NIR and IR luminescence as well. All Nd-doped samples showed emission in the NIR (~1060nm and ~1350nm) while the Er-doped samples' emission was weak at ~1546 nm. Substrate conditions play a significant role in magnetic behavior as both Si and RE atoms compete for gallium sites, and increased Si-doping leads to decreased Ms. Ms increases and saturates with longer annealing times. Luminescent intensity varies slightly over the Si doping range, and emission intensity also varies with annealing time. To the best of our knowledge, this is the first demonstration of above RT ferromagnetism and emission in GaNdN via diffusion.

REFERENCES

1. M. L. Reed, N. A. El-Masry, H. H. Stadelmaier, M. K. Ritums, M. J. Reed, C. A. Parker, J. C. Roberts, and S. M. Bedair, Appl. Phys. Lett., 79, 3473 (2001).
2. S. Sonoda, S. Shimizu et al., J. Cryst. Growth, 237-239, 1358 (2002).
3. J. M. Zavada, H. Nepal et al., Appl. Phys. Lett., 91, 054106 (2007).
4. Y-S. Ting, C-C. Chen et al., Optical Materials, 24, 515-518 (2003).
5. C-C. Chen, Y-S. Ting et al., Solid-State Electronics, 47, 529-531 (2003).
6. S. J. Pearton, H. Cho et al., Appl. Phys. Lett., 75, 2939 (1999).
7. I. D. Hughes, M. Däne et al., Nature, 446, 650-653 (2007).
8. H. Bang, J. Sawahata et al., Phys. Stat. Sol. (c), 0 (7), 2874-2877 (2003).
9. C. M. Aerts, P. Strange et al., Phys. Rev. B, 69, 045115 (2004).
10. C. G. Van de Walle and J. Neugebauer, Brazilian Journal of Physics 26, 163 (1996).
11. T. Dietl, Physica E, 35 (2006) 293-299.
12. A. J. Steckl, J. C. Heikenfeld, D. S. Lee, M. Garter, C. C. Baker, Y. Q. Wang and R. Jones, IEEE J. Select. Top. Quantum Electron., 8, 749-766 (2002).
13. J. H. Kim and P. H. Holloway, Adv. Mater., 17, 91-96 (2005).
14. D. M. Hansen, R. Zhang et al., Appl. Phys. Lett. 72, 1244 (1998).
15. R. Birkhahn and A. J. Steckl, Appl. Phys. Lett. 73, 2143 (1998).
16. J. Devin MacKenzie, C. R. Abernathy, S. J. Pearton, U. Hömmerich, J. T. Seo, and R. G. Wilson, Appl. Phys. Lett., 72 2710 (1998).
17. D. S. Lee, J. Heikenfeld et al., Appl. Phys. Lett., 79, 719 (2001).
18. E. D. Readinger, G. D. Metcalfe et al., Appl. Phys. Lett., 92 061108 (2008).

Mater. Res. Soc. Symp. Proc. Vol. 1183 © 2009 Materials Research Society 1183-FF06-04

Novel room-temperature ferromagnetic semiconductors Pb$_{1-x-y}$Ca$_x$Cr$_y$Te

Evgeny P. Skipetrov[1], Elena A. Zvereva[1], Nikolay A. Pichugin[1], Alexey E. Primenko[1], Evgeny I. Slyn'ko[2] and Vasily E. Slyn'ko[2]

[1]Faculty of Physics, Moscow State University, Moscow, 119992, Russia
[2]Institute of Material Science Problems, Chernovtsy, 274001, Ukraine

ABSTRACT

The galvanomagnetic and magnetic properties of novel diluted magnetic semiconductors Pb$_{1-x-y}$Ca$_x$Cr$_y$Te (x=0.06-0.20, y=0.003-0.045) have been investigated. Temperature dependencies of the resistivity and the Hall coefficient have a metallic character indicating the pinning of Fermi level by the chromium impurity level on the background of the conduction band states. Magnetization curves display a clear hysteresis loop over the whole temperature range investigated. The Curie temperature, determined from the temperature dependencies of magnetization, achieves 345 K. Possible mechanisms of ferromagnetic ordering were discussed.

INTRODUCTION

Ferromagnetic diluted magnetic semiconductors (DMSs), which combine semiconductor properties and ferromagnetism, are in focus of a great interest in recent years because of their important applications in the field of spintronics. One of the major challenges is to obtain a DMS with the Curie temperature T$_C$ exceeding the room temperature. So far, most part of the efforts has been conducted on group III-V manganese-doped DMSs [1-3]. Generally, it has been assumed, that long-range ferromagnetic interactions in these materials are mediated by valence-band holes in accordance to the Ruderman-Kittel-Kasuya-Yosida (RKKY) model. However, no significant progress has been made for the last four years since the highest reported value of the Curie temperature T$_C$=173 K for (Ga,Mn)As [4], which is still far below the room temperature.

More recently, in accordance with the theoretical predictions [5, 6], the room-temperature ferromagnetism has been revealed in the wide-gap n-type DMSs GaN(Cr) (T$_C$ up to 280 K) [7], Zn$_{1-x}$Cr$_x$Te (T$_C$ up to 300 K) [8], Cd$_{1-x}$Cr$_x$Te (T$_C$ up to 395 K) [9] and even in narrow-gap IV-VI DMSs Pb$_{1-x-y}$Ge$_x$Cr$_y$Te (T$_C$ up to 280 K) [10]. However, the origin of ferromagnetism for these systems cannot be explained by the carrier-mediated RKKY mechanism.

In the present work we report on the galvanomagnetic and magnetic properties of novel chromium doped DMSs Pb$_{1-x-y}$Ca$_x$Cr$_y$Te. The main objectives were to reveal a new room-temperature DMS, to obtain the energy level diagram for these alloys and to find a connection between magnetic and galvanomagnetic properties under variation of the matrix composition and magnetic impurity content. On the contrary to Pb$_{1-x}$Ge$_x$Te alloys, which are known to undergo the structural phase transition from cubic to rhombohedral phase at lowering temperature [11], Pb$_{1-x}$Ca$_x$Te remain cubic crystals over the whole range of the alloy composition x with relatively small variation of the lattice constant [12, 13]. So, one can expect essential rising of the crystal perfection and, hence, uniformity of the physical properties of these alloys. Besides, in spite of smaller atomic radius of calcium, it easy incorporates into the cubic PbTe semiconductor lattice substituting the lead positions and may enhance the solubility of magnetic chromium impurity.

EXPERIMENTAL TECHNIQUES

A single crystal ingot of $Pb_{1-x-y}Ca_xCr_yTe$ has been grown by the Bridgman method. The ingot was cut into slices approximately 1.5 mm thick perpendicular to the growth direction. The chemical composition was determined using the X-ray fluorescence analysis. It was found that the calcium concentration monotonously decreases (x=0.06-0.20), while the chromium content grows (y=0.0.003-0.045) at moving from the origin to the end of the ingot in accordance with the typical for group IV-VI solid solutions exponential distribution of substitution components [14]. The incorporation of calcium and chromium into the host semiconductor lattice has been confirmed by the structural characterization. Powder high-resolution X-ray diffraction (XRD) studies were carried out at room temperature with CuK_α radiation source on Rigaku X-ray diffractometer with monochromator. The diffraction pattern displays a good matching with PbTe cubic structure with no second phases of the ferromagnetic compounds at least up to y=0.025.

For galvanomagnetic measurements the specimens, approximately $3\times0.7\times0.7$ mm^3 in size, were cut from the original crystal slices using an arc cutting machine. Voltage and Hall contacts, made of platinum wire, were attached by spark-discharge welding to the samples. Current contacts were soldered to the samples using an alloy of indium with 4%Ag and 1%Au. For each sample the temperature dependencies of the resistivity ρ and Hall coefficient R_H (B≤0.08 T, 5≤T≤300 K) and magnetic field dependencies of Hall coefficient R_H (T=4.2 K, B<8 T) have been measured by standard four-probe technique.

Magnetic measurements were performed on the samples of the weight 0.05-0.1 g glued onto a plastic holder. The temperature and magnetic field dependencies of the magnetization were measured using vibrating sample magnetometer EG & G PARC M155 detecting the magnetic moments down to 5×10^{-5} emu. The magnetic field up to B=0.5 T was produced by an electromagnet and determined by means of Hall sensor. For the temperature variation (5≤T≤300 K) the liquid helium gas-flow cryostat equipped with a heater was used.

RESULTS AND DISCUSSION

Galvanomagnetic properties

It has been found that all investigated samples have the n-type conductivity. Temperature dependencies of resistivity $\rho(T)$ and the Hall coefficient $R_H(T)$ demonstrates a metal-like behavior (figure 1). However, the anomalous increase of R_H absolute value with the increase of temperature is observed. Such a behavior of the Hall coefficient has been observed earlier in the lead telluride-based alloys doped with mixed-valence impurities (In, Ga, Cr, etc.) and attributed to the pinning of the Fermi level by the resonant impurity level, situated in the conduction band. At helium temperatures resistivity monotonously rises with the increase of the calcium concentration and the decrease of magnetic Cr impurity in the alloys inducing the decrease of the Hall mobility approximately three times. The Hall coefficient only slightly deviates from the value $R_H=-(0.4-0.5)$ cm^3/C in the whole investigated ranges of composition.

Using the values of the Hall coefficient at helium temperature the dependence of the free electron concentration on the calcium content n(x) was determined. Then in the frame of two-band Kane dispersion relation for A^4B^6 semiconductors [11] the Fermi energy versus alloy composition dependence $E_F(x)$ were calculated. And finally, assuming the pinning of the Fermi

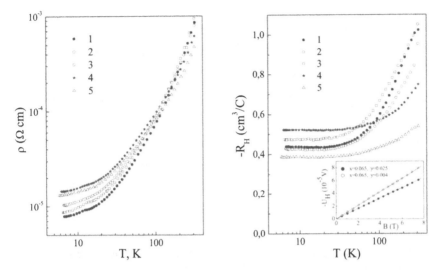

Figure 1. Temperature dependencies of resistivity and Hall coefficient in $Pb_{1-x-y}Ca_xCr_yTe$ (x, y: 1 – 0.060, 0.006; 2 – 0.075, 0.003; 3 – 0.095, 0.003; 4 – 0.130, 0.003; 5 – 0.160, 0.003). In the inset – magnetic field dependencies of the Hall voltage.

level in the investigated alloys by the chromium deep level and an increase of the gap in $Pb_{1-x}Ca_xTe$ alloys with the initial rate $dE_g/dx \approx 25$ meV/mol.% [12, 13], we have obtained the diagram of impurity level movement relative to the principal band edges for our $Pb_{1-x-y}Ca_xCr_yTe$ alloys (figure 2). In order to expand the range of the alloy composition under consideration a well-known data for the electron concentration and Fermi level position in the chromium doped PbTe were added in the n(x) and $E_F(x)$ dependences [15, 16].

It was shown that electron concentration passes through the maximum at $x \approx 0.05$ and tends to zero with the increase of the calcium content. The chromium impurity level moves almost linearly relative to the conduction band bottom with the rate $d(E_{Cr}-E_c)/dx \approx -4$ meV/mol.%. So one can conclude, that increase of the calcium content should lead to the interception of chromium deep level with the conduction band bottom and transition to the insulator phase at $x \approx 0.25$. Essential deviation of experimental points for the alloys with maximal calcium concentration from the theoretical line in the energy level diagram (see figure 2) may be caused by the nonlinearity of the composition dependence of the gap, experimentally revealed for $Pb_{1-x}Ca_xTe$ alloys with x>0.15 [12, 13], and by the absence of the Fermi level pinning in the samples with minimal impurity content.

Magnetic properties

Magnetization curves display a clear hysteresis loops over the whole temperature range investigated (figure 3). The coercive force passes through the maximum with increasing of chromium concentration and achieves 0.13 T at 5 K. Magnetic saturation moment increases

53

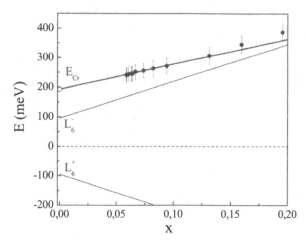

Figure 2. Energy level diagram for $Pb_{1-x-y}Ca_xCr_yTe$ alloys under variation of the matrix composition (open point – previously known data for chromium doped PbTe [15, 16]).

rapidly with increasing chromium concentration in the alloys. Assuming that the magnetic moment arises solely from Cr impurity atoms the effective magnetic moment per chromium atom has been estimated. It was found to be about 0.6-0.7 μ_B in the middle part of the ingot and increases up to 1.5-2 μ_B at the ends of the crystal ingot. This values are lower than the moment per Cr atom in the known magnetic chromium compounds, for example, CrTe (~2.6 μ_B) [16], Cr_3Te_4 (~2.0 μ_B) [17], Cr_5Te_8 (~3.8 μ_B) [18] and satisfactory agrees with the values, obtained for Cr-doped DMSs GaN(Cr) (~1.1 μ_B) [7], $Zn_{1-x}Cr_xTe$ (~0.96 μ_B) [8].

The temperature dependencies of the magnetization M(T) is rather concave type with a broaden maximum at temperatures about 130-150 K for all samples under study (see figure 3). The amplitude of this maximum decreases with decreasing the magnetic field applied. On the temperature dependencies of residual magnetization $M_r(T)$ this maximum disappears and we observed a plateau at lower temperatures. Absolute value of residual magnetization at T=5 K increases from 0.05 to 2.5 emu/g with increasing chromium content in the alloys. We have observed also the difference between ZFC and FC magnetization curves for the alloys with low Cr concentration up to relatively high freezing temperature $T_f \approx 120$ K, that may indicate the existence of spinglass phase at low temperatures. Both with an increase in chromium concentration and in magnetic field T_f shifts towards the lower temperature and finally ZFC curve merges with FC one.

The Curie temperature was obtained by extrapolating of the linear part of M^2 versus temperature dependence to its temperature intercept. It was found that the ferromagnetic ordering temperature weakly increases with increasing chromium concentration in the alloys and its maximal value achieves T_C=345 K (see inset in figure 3).

The mechanism of the ferromagnetic ordering in our alloys is unclear at the moment. Low concentration of electrons ($n \approx 10^{19}$ cm^{-3}) and weak impurity concentration dependence of the Curie temperature $T_C(y)$ make us to conclude that the carrier-mediated RKKY mechanism can

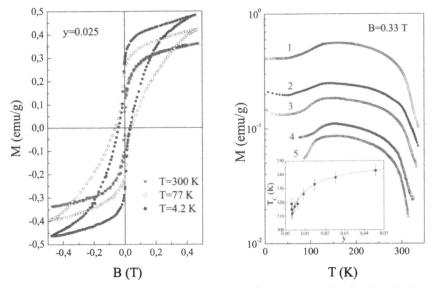

Figure 3. . Hysteresis loops and temperature dependencies of magnetization for Pb$_{1-x-y}$Ca$_x$Cr$_y$Te (x≈0.07). y: 1–0.025, 2–0.015, 3–0.010, 4–0.006, 5–0.005.

not explain ferromagnetism in our samples. On the basis of our XRD and magnetisation data one can suppose intrinsic character of ferromagnetism rather than influence of ferromagnetic secondary phases in the alloys. Besides, high electron mobility (up to 10^5 cm^2/Vs) and quantum oscillations of magnetoresistivity (Shubnikov-de Haas effect) at helium temperatures were observed, which may be considered as additional confirmation of this conclusion. If so, the superexchange interaction between Cr ions, which was suggested for Cr-doped II-VI DMSs [20, 8], might play a leading role in our case too. Another possible explanation of ferromagnetic ordering is a new mechanism based on the exchange coupling of the band states and localized spins and on the hybridization of the band states with resonant chromium impurity states [21].

CONCLUSIONS

The chromium-induced deep level located on the background of the conduction band of Pb$_{1-x-y}$Ca$_x$Cr$_y$Te (0.06≤x≤0.20) has been revealed. In the frame of the two-band Kane dispersion relation the composition dependencies of the electron concentration and Fermi energy in the alloys have been obtained. The energy level diagram as a function of the matrix composition has been constructed. It was found that with increasing of calcium content Cr resonant level almost linearly moves up relative to the middle of the gap, approaching the conduction band bottom with the rate d(E$_{Cr}$-E$_c$)/dx≈-4 meV/mol.%.

The room-temperature ferromagnetism in the alloys investigated was revealed. It was shown, that Curie temperature, obtained by extrapolating of the linear part of M^2 versus temperature dependencies to its temperature intercept, weakly grows with the increase of chromium content and achieves T_C=345 K at y≈0.045. Possible mechanism of ferromagnetic ordering were discussed involving the superexchange interaction between chromium ions and the exchange coupling of the band states and localized spins in the case of the hybridization of the band states with resonant impurity states.

ACKNOWLEDGMENTS

The authors would like to thank Dr. S.A. Ibragimov for performing the X-ray diffraction measurements. This work has been supported by Russian Foundation for Basic Research (Grant No. 08-02-01364).

REFERENCES

1. T. Dietl, H. Ohno, Mater. Today, 9, 18 (2006).
2. S. Kuroda, N. Nishizawa, K. Takita, M. Mitome, Y. Bando, K. Osuch, T. Dietl, Nature Mater., 6, 440 (2007).
3. T. Dietl, J. Appl. Phys. 103, 07D111 (2008).
4. K. Wang, K. Edmonds, R. Champion, B. Gallagher, T. Foxon, M. Sawicki, T. Dietl, P. Boguslawski, and T. Jungwirth, *Proceedings of the 27th International Conference on Physics of Semiconductors*, edited by J. Mendez, (AIP, Melville, 2005), p.333.
5. T. Dietl, H. Ohno, F. Matsukara, J. Cilbert, and D. Ferrand, Science, 287, 1019 (2000).
6. T. Dietl, H. Ohno, and F. Matsukara, Phys. Rev. B, 63, 195205 (2001).
7. S.E. Park, H.-J. Lee, Y.C. Cho, S-Y. Jeong, C.R. Cho, S. Cho, Appl. Phys. Lett., 80, 4187 (2002).
8. H. Saito, V. Zayets, S. Yamagata, K. Ando, Phys. Rev. Lett., 90, 207202 (2003).
9. K.Y. Ko, M.G. Blamire, Appl.Phys. Lett., 88, 172101 (2006).
10. E.P. Skipetrov, M.G. Mikheev, F.A. Pakpour, L.A. Skipetrova, N.A. Pichugin, E.I. Slyn'ko, and V.E. Slyn'ko, Semiconductors, 43, 297 (2009).
11. R. Dornhaus, G. Nimtz and B. Schlicht, Narrow-Gap Semiconductors (Springer-Verlag, Berlin, 1983).
12. D.L. Partin, IEEE J. Quantum Electronics, 24, 1716 (1988).
13. A. Ishida, T. Tsuchiya, N. Yoshioka, K. Ishino, and H. Fujiyasu, Jpn. J. Appl. Phys., 38, 4652 (1999).
14. V.E. Slyn'ko, Visnyk Lviv Univ., Ser. Physic. 34, 291 (2001).
15. V.D. Vulchev, L.D. Borisova, and S.K. Dimitrova, phys. stat. sol. (a), 97, K79 (1986).
16. V.D. Vulchev, and L.D. Borisova, phys. stat. sol. (a), 99, K53 (1987).
17. A. Ohsawa, Y. Yamaguchi, N. Kazama, H. Yamauchi, and H. Watanabe, J. Phys. Soc. Jap., 33, 1303 (1972).
18. M. Yamaguchi, T. Hashimoto, J. Phys. Soc. Jap., 32, 635 (1972).
19. K. Lukoschus, S. Krashinski, C. Nather, W. Bensch, and R.K. Kremer, J. Sol. St. Chem., 177, 951 (2004).
20. J. Blinowski, P. Kacman and J.A. Majewski, Phys. Rev. B53, 9524 (1996).
21. V.K. Dugaev, V.I. Litvinov, J. Barnas, A.H. Slobodskyy, W. Dobrowolski and M. Vieira, J. Supercond.: Incorporated Novel Magnetism 16, 67 (2003).

Mater. Res. Soc. Symp. Proc. Vol. 1183 © 2009 Materials Research Society   1183-FF06-05

Spin polarization of electrons injected from Fe into GaAs quantum well characterized using oblique Hanle effect

Eiji Wada[1], Mitsuru Itoh[1], Tomoyasu Taniyama[1,2] and Masahito Yamaguchi[3]
[1]Materials and Structures Laboratory, Tokyo Institute of Technology,
4259 Nagatsuta, Midori-ku, Yokohama 226-8503, Japan
[2]PRESTO-JST, 3-5, Sanbancho, Chiyoda-ku, Tokyo 102-0075, Japan
[3]Graduate School of Engineering, Nagoya University,
Furo-cho, Chikusa-ku, Nagoya 464-8603, Japan

ABSTRACT

We study spin injection from an in-plane magnetized Fe thin layer into a GaAs/AlGaAs quantum well (QW) in low magnetic fields of ±0.37 T using oblique Hanle effect. An oblique low magnetic field induces the precession of electron spins in the GaAs QW, allowing us to detect the spin polarization of electrons injected across the Fe/AlGaAs interface. Our analysis of the circular polarization of light emitted in the electron-hole recombination process in the QW gives an estimate of the lower bounds of the spin polarization to be 4.0%. Also, a spin lifetime of 140 psec is obtained in this analysis, indicating that spin depolarization at the Fe/AlGaAs interface is more predominant rather than spin relaxation in the QW region.

INTRODUCTION

A prerequisite for realizing spintronic devices, e.g., spin transistors, is to inject electron spins efficiently into a III-V semiconductor such as GaAs. Most recent studies of spin injection into GaAs use spin light emitting diode (spin-LED) structures to optically estimate the spin polarization of electrons injected. A spin-LED structure principally consists of a ferromagnetic thin layer and a semiconductor quantum well (QW), where spin polarized electrons are injected from the ferromagnetic layer into the QW active region, thereby light emission occurs. According to the optical selection rules, the optical circular polarization of the emitted light is directly related to the spin polarization of electrons injected, therefore, a very reliable estimate of the spin polarization can be obtained by analyzing the circular polarization of the light emission. However, this approach requires large magnetic fields of 2-3 Tesla to saturate the magnetization of a ferromagnetic spin injector normal to the layer plane, which is incompatible with practical spintronic devices, since the circular polarization of the light is detected from the direction

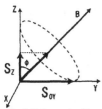

Fig.1 Schematic diagram of Larmor precession of an electron spin. Out-of-plane component of the spin appears.

perpendicular to the ferromagnetic layer plane (Faraday geometry)[1-3]. Oblique Hanle effect (OHE), on the other hand, allows us to circumvent the use of large magnetic fields to estimate the spin polarization. In the OHE geometry, a low magnetic field of the order of kOe applied at an angle ϕ from normal to the layer plane induces Larmor precession of an electron spin about the magnetic field in the semiconductor, thereby the net out-of-plane component of the spin appears even using an in-plane magnetized ferromagnetic spin injector (see Fig.1). Therefore, OHE is a very effective means to estimate the spin polarization of electrons injected from an in-plane magnetized ferromagnetic layer into a semiconductor QW even in low magnetic fields.

In this study, we report on spin injection from an epitaxial Fe layer into a GaAs/AlGaAs QW in the OHE approach. Our choice of Fe as a spin injector is based on the fact that the lattice constant of Fe (2.8 Å) well matches with that of AlGaAs, enabling us to grow well defined epitaxial Fe layers on AlGaAs. Although spin injection from a ferromagnetic metal into a semiconductor encounters an obstacle due to conductivity mismatch, which decreases spin injection efficiency in the diffusive transport limit[4], tunneling spin transport across the Schottky barrier at Fe/AlGaAs likely overcomes the difficulty due to the conductivity mismatch as demonstrated in a previous work on spin injection from Fe into GaAs under a large magnetic field at room temperature[5].

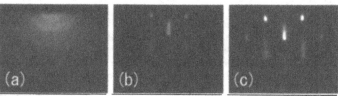

Fig.2 RHEED patterns of (a) an As capping layer, (b) an n-AlGaAs surface, and (c) an Fe layer.

EXPERIMENT

Fe/GaAs/AlGaAs QW heterostructures consisting of a 15-nm-thick Fe layer and a 20-nm-thick GaAs QW were fabricated in ultrahigh vacuum MBE chambers with a base pressure of 10^{-10} Torr. The active region of the GaAs QW structure has a 75-nm-thick n-Al$_{0.1}$Ga$_{0.9}$As (1×10^{18} cm^{-3}), a 20-nm-thick undoped GaAs QW, and a 25-nm-thick p-Al$_{0.3}$Ga$_{0.7}$As (1×10^{18} cm^{-3}), with an As capping layer. The As capped QW structure was transferred to another MBE chamber and the As layer was removed by annealing at 480 °C for 15 min, prior to the growth of the Fe layer. After checking the flat AlGaAs surface in the reflection high energy electron diffraction (RHEED) pattern, the growth of an Fe layer was carried out at room temperature at a growth rate of 0.1 Å/sec. Figure 2 shows the RHEED patterns of each step during the growth. The RHEED from the Fe surface shows a streaky pattern, ensuring that the epitaxial growth of the Fe layer occurs on the AlGaAs surface.

Electroluminescence (EL) spectra from the Fe/GaAs/AlGaAs QW heterostructure were measured at an energy range of 1.47-1.52 eV using a monochromator and a charge coupled device array. Circular polarization analysis of the emitted light was done using a quarter-wave plate and a linear polarizer. A magnetic field of ±0.37 T was applied along the direction 45 degrees away from normal to the layer plane for the circular polarization analysis, so as to maximize the net out-of-plane component of electron spin due to the oblique Hanle effect. Emitted light with either right or left circular polarization was collected separately by rotating the quarter-wave plate. All the measurements were done at 80 K in an optical cryostat.

DISCUSSION

Figure 3 shows the EL spectra measured at 80 K under a bias voltage of 2.8V for -0.37 T, 0 T, and 0.37 T. A clear peak due to electron-hole excitons in a QW is seen at 1.49 eV, accompanied by a small shoulder at 1.51 eV. The small shoulder is not likely associated with the splitting of the heavy hole (HH) and light hole (LH) subbands of GaAs since the energy difference between the peak and the shoulder is too large compared with the HH-LH splitting. Although the origin of the shoulder is not clear yet, some transition processes might contribute to the shoulder in the EL spectra. We note that the EL peak intensities corresponding to the right and left circular polarization are different to each other in a magnetic field and the relative magnitude of the intensities switches by reversing the magnetic field, clearly indicating that the spin polarized electrons injected from the in-plane magnetized Fe layer are detected in the geometry of OHE.

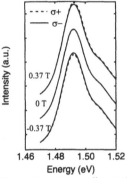

Fig.3 EL spectra corresponding to the right ($\sigma+$) and left ($\sigma-$) circular polarization in magnetic fields of 0.37 T, 0 T, and -0.37 T at 80 K.

In the OHE geometry, the out-of-plane component of electron spin, $S_z(B)$, is given by eq.(1) in the steady state, as derived from the Bloch-type equation[6].

$$S_z = \frac{1}{2}S_{0y}\frac{T_S}{\tau}\frac{(B/\Delta B)^2}{1+(B/\Delta B)^2} , \qquad (1)$$

where S_{0y}, τ and T_S are the in-plane component of electron spin injected just underneath the Fe/AlGaAs interface, the electron recombination lifetime, and the spin lifetime ($T_S^{-1}=\tau^{-1}+ \tau_S^{-1}$, where τ_S is the spin scattering time), respectively. Also, ΔB is given as

$$\Delta B = (\frac{g^* \mu_B}{\hbar} T_S)^{-1}, \tag{2}$$

where g^* is the effective g factor, $\hbar$ and μ_B are the Planck constant and the Bohr magneton, respectively. Although electrons in transmitting through the n-AlGaAs from the Fe layer to the QW might contribute to g^*, the value of g^* is likely arising from electrons in the QW, since the transmission time of an electron in the AlGaAs layer is a factor of 10^5 smaller than the electron-hole recombination time in the QW[7]. Therefore, the value g^* of 0.43 for a 20-nm-thick GaAs QW is a reasonable choice for this case[8]. Also, the spin polarization of injected electrons P_{spin} is given as $2S_{0y}$[6], and P_{spin} is estimated to be $P_{\text{spin}}=2S_{0y}=4S_z^{\text{max}} \tau/T_S$ using the maximum value S_z^{max} of S_Z in eq.(1)[6]. The circular polarization P_{circ} of EL, on the other hand, is given as $P_{\text{circ}}=S_z$, since the HH and LH are degenerated in the QW. Therefore, by fitting the field dependence of P_{circ} with eq.(1), we can obtain an estimate of the parameter $S_z^{\text{max}} =P_{\text{circ}}^{\text{max}}=P_{\text{spin}}T_S/4\tau$, in other words, the spin polarization of electrons injected can be given as $P_{\text{spin}}=4S_z^{\text{max}} \tau/T_S$.

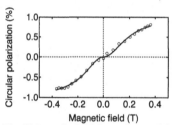

Fig.4 Magnetic field dependence of the circular polarization of emitted light. A fitted curve with eq. (1) is represented by a solid curve.

Now, we discuss the magnetic field B dependence of the circular polarization of EL spectra. Figure 4 demonstrates the magnetic field dependence of the circular polarization P_{circ}. Eq. (1) is fitted to the magnetic field dependence, yielding ΔB=0.187 and S_z^{max}=1.0%, and we obtain the spin polarization of electrons injected P_{spin}=4.0%× τ/Ts, accordingly. Since the value of τ/T_S is larger than unity[6], the spin polarization should be greater than 4.0% as the lower bounds at 80 K. Also, the value of ΔB gives T_S=140 psec using eq. (2). In order to precisely estimate the degree of spin polarization, time resolved photoluminescence measurement is required for the evaluation of the recombination time τ. We should note that Zeeman effect in a QW and magnetic circular dichroism (MCD) could contribute to the circular polarization of the emitted light as an artifact. However, these effects are not significant in an oblique low magnetic field ~0.3 T, compared to the out-of-plane magnetic field larger than 1 T used in a Faraday geometry. In fact, the out-of-plane spin polarization is estimated to be 0.095 % at an oblique magnetic field of 0.37 T using the spin polarization P_z due to Zeeman effect given by

$$P_z = \tanh(\frac{g^* \mu_B B}{k_B T}),$$

where k_B is the Boltzmann constant and g^* is 0.43, giving rise to a circular polarization of 0.0425 %. The value is negligibly small compared to the result in Fig.4. Therefore, the predominant contribution in Fig. 4 likely has its origin in the spin injection across the Fe/AlGaAs interface.

The value of the spin polarization estimated in this study is not large compared with that in a previous report on a CoFe/AlO$_X$ injector, where a spin polarization of 21% was reported[6]. The value of T_S, on the other hand, is comparable to those reported for Fe/InGaAs QW and Co$_{2.4}$Mn$_{1.6}$Ga/InGaAs structures, where large spin polarization of 31% and 13% were estimated at 5 K, respectively[9]. These combined results indicate that the low spin polarization observed in this study is due not to spin relaxation in the GaAs QW but to spin-flip scattering at the Fe/AlGaAs interface, presumably due to the formation of interdiffusion layer such as Fe-As compound, although detailed analysis of the interface is now in progress.

CONCLUSIONS

We have reported spin injection from an in-plane magnetized Fe layer into a GaAs quantum well by using oblique Hanle effect. The spin polarization of electrons was estimated to be 4.0% as the lower bounds and the spin lifetime of 140 psec was obtained. In spite of the low spin polarization, the large spin lifetime likely indicates that spin depolarization occurs due to spin-flip scattering at the Fe/AlGaAs interface.

ACKNOWLEDGMENTS

This work was supported in part by KAKENHI (18686050) from The Ministry of Education, Culture, Sports, Science and Technology (MEXT) and Yoshida Foundation for Science and Technology.

REFERENCES

1. B. T. Jonker, Y. D. Park, B. R. Bennett, H. D. Cheong, G. Kioseoglou, and A. Petrou, Phys. Rev. B, **62**, 8180 (2000).
2. A. Kawaharazuka, M. Ramsteiner, J. Herfort, H. -P. Schönherr, H. Kostial, and K. H. Ploog, Appl. Phys. Lett., **85**, 3492 (2004).
3. X. Jiang, R. Wang, R. M. Shelby, R. M. Macfarlane, S. R. Bank, J. S. Harris and S. S. P. Parkin, Phys. Rev. Lett., **94**, 056601 (2005).
4. G. Schmidt, D. Ferrand, L. W. Molenkamp, A. T. Filip, and B. J. van Wees, Phys. Rev. B, **62**, R4790 (2000).
5. H. J. Zhu, M. Ramsteiner, H. Kostial, M. Wassermeier, H. –P. Schönherr, and K. H. Ploog, Phys. Rev. Lett., **87**, 016601 (2001).
6. V. F. Motsnyi, P. Van Dorpe, W. Van Roy, E. Goovaerts, G. Borghs, and J. De Boeck, Phys. Rev. B, **68**, 245319 (2003).

7. M. Gurioli, A. Vinattieri, and M. Colocci, C. Deparis, J. Massies, G. Neu, A. Bosacchi, and S. Franchi, Phys. Rev. B, **44**, 3115 (1991).
8. M. J. Snelling, E. Blackwood, C. J. McDonagh, R. T. Harley, and C. T. B. Foxon, Phys. Rev. Lett., **45**, 3922 (1992).
9. M. C. Hickey, C. D. Damsgaard, I. Farrer, S. N. Holmes, A. Husmann, J. B. Hansen, C. S. Jacobsen, D. A. Ritchie, R. F. Lee, G. A. C. Jones, and M. Pepper, Appl. Phys. Lett., **86**, 252106 (2005).

Mater. Res. Soc. Symp. Proc. Vol. 1183 © 2009 Materials Research Society   1183-FF06-14

Temperature-Dependent Structural Disintegration of Delafossite CuFeO$_2$

Shojan P. Pavunny, Ashok Kumar, R. Thomas, N.M. Murari and R.S. Katiyar

Department of Physics and Institute for Functional Nanomaterials, University of Puerto Rico, P.O. Box 23343, San Juan, PR, USA

ABSTRACT

Single phase delafossite p-type CuFeO$_2$ (CFO) semiconductor was synthesized in bulk by modified solid state reaction technique. X-ray diffraction (XRD) and X-ray photo spectroscopy (XPS) studies suggest single phase CFO at room temperature. The energy dispersive X-ray spectroscopy (EDX) revealed that the atomic ratio of Cu and Fe is 1:1. The XPS spectra showed two intense Cu $2p_{3/2}$ and $2p_{1/2}$ peaks at 932.5 eV and 952 eV suggesting Cu is in +1 state. The temperature dependent Raman spectra of CFO displayed two intense modes at 349 cm^{-1} and 690 cm^{-1} at room temperature that matched with other delaffosite structures. The temperature dependent Raman spectra showed significant shift in both Raman active modes to lower frequency side. We observed the disappearance of pure CFO Raman active modes above 750 K and the appearance of new peaks related to CuO compounds, indicating disintegration of CFO starting above 750 K which almost completed above 1100 K. The temperature dependent thermo-gravimetric analysis indicates change in CFO mass above 750 K with wide range of differential thermo-gravimetric slope suggests disintegration started above 750 K and completed at 1100 K. Raman spectra, XPS, and XRD of disintegrated CFO matched well with the Raman spectra, XPS and XRD of CuO and CuFe$_2$O$_4$ confirmed its disintegration above 750 K in air.

INTRODUCTION

Delafossite, a type of ternary oxides has the general formula ABO$_2$ where, A cations are linearly coordinated with two oxygen ions and B cations are situated in the distorted edge-shared BO$_6$ octahedra [1-2]. Friedel [3] realized the existence of CFO in 1873. The CFO structure consists of hexagonal layers of Cu, O, and Fe with a stacking sequence of Cu-O-Fe along the c-axis to form a layered triangular lattice antiferromagnet. The preparation of such compounds by the solid state reaction is a challenge as these have the tendency to decompose before the formation reaction occurs. CFO has versatile functional properties i.e. superconductivity, large magneto resistance, thermoelectric effects, and multiferroicity that makes them potential candidates for device applications .It showed several interesting antiferromagnetic phase transitions between 4 K to 50 K due to geometrical frustration at low temperature [4-7].

Surprisingly, there is no report on Raman Spectroscopy studies on polycrystalline ceramics, thin films, or single crystals of CFO at the ambient temperature. There are a few reports on pressure and temperature dependent lattice dynamics of delafossite CuGaO$_2$ and CuAlO$_2$, which were investigated by Raman spectroscopy [8-10]. Pellicer-Porres et al. [10] observed two Raman active modes for CuAlO$_2$ at Eg ~ 418 cm^{-1} and Ag ~ 767 cm^{-1}, similar to other wurtzite structure [11]. Pellicer-Porres et al. [10] discussed temperature dependent phonon dispersion behavior in CuAlO$_2$ and showed linear dependence up to 200 K; above 200 K, a deviation in the phonon energy was observed which was explained on the basis of the thermal expansion. We report the synthesis of single phase delafossite CFO and its temperature dependent decomposition in open atmosphere well probed by in situ Raman Spectroscopy and confirmed by XRD, XPS and thermo-gravimetric analysis. The lattice dynamic behavior of single phase CFO and disintegrated CFO was probed by micro Raman spectroscopy at ambient temperature and high temperature. As far as we know, no prior studies were carried out on Raman spectroscopy of CFO.

EXPERIMENT

Polycrystalline CFO was prepared by high energy solid state reaction from a stoichiometric mixture of Cu_2O and Fe_2O_3 powders. This mixture was high energy ball milled for 1 hour and then calcined at 950°C for 12hrs and at 1000°C for 12hrs using Carbolite HTF1700 furnace. During calcinations, the alumina crucible automatically sealed with the alumina cover. This may be due to internally generated inward pressure. The as-synthesized CFO compound showed Delafossite phase in XRD measurement (Siemens D5000 XRD). Calcined powder was pressed in the form of thick pellet at 4Ton pressure and later sintered in Argon atmosphere at 950°C for 4hrs. XRD spectra of the sintered pellets proved that the delafossite phase is preserved within the experimental limit.

Energy-dispersive X-ray (EDX) spectra of the pellet were recorded to identify the elements present in the sample. Microstructure analysis was carried by Jeol Scanning Electron Microscope (SEM) at 3000x, 10000x and 25000x magnifications. X-ray photoelectron spectroscopy (XPS) was employed to study the elemental composition, empirical formula, chemical state and electronic state of the elements that exist in the compound.

Delafossite structural analysis and its thermal decay study were carried out in air by employing temperature dependent Raman Spectroscopy using Jobin Yvon T64000 spectrometer. The laser line at 514.5nm from a Coherent Argon ion laser, Innova 90-5 was focused on the sample. A liquid Nitrogen cooled CCD device collected the Raman scattered signal through a 50X objective. High temperature Raman spectra of the sample were collected in air from 300 K to 1000 K using Linkam module. Raman spectra were also recorded for the same sample once it was cooled down to room temperature. The Thermal Gravimetric Analysis (TGA) of the as-synthesized samples were carried out in a Shimadsu DTA50. The samples were heated from room temperature up to 1000°C at the rate of 5 °C/min in air. The Differential Thermal Analysis (DTA) was conducted with the same above mentioned parameters.

RESULTS AND DISCUSSION

Figure 1(a) shows the XRD pattern of CFO in the form of powder and pellet. All peaks are indexed using computer controlled POWD program [12] according to the hexagonal delafossite structure with $R\bar{3}m$ space group well consistent with the literature values (JCPDS). No additional/impurity peaks were evidenced, showing the good quality of the synthesized powder. The calculated lattice constants were a = 3.03 Å and c = 17.14 Å at room temperature matched well with the earlier report and JCPDS files [13]. Figure 1(b)(inset) represents the SEM image of sintered CFO pellet. The well defined layered granular grains have an average size ~ 5 μm. Around 10 to 15% porosity were observed in pellets due to low sintering temperature. The corresponding energy dispersive spectra (EDX) showed that average atomic ratio of Cu: Fe is

1:1. This value of atomic ratio sustains the existence of Cu and Fe in +1 and +3 electronic state respectively in the delafossite structure. Further investigations were carried out to check the

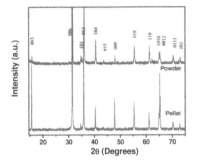 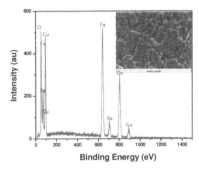

Figure 1(a). XRD pattern of CFO powder and pellet at room temperature and **(b)** EDX spectrum and SEM image (inset) of CFO pellet.

chemical state of Cu and Fe, and their distribution in the as-synthesized CFO powder by XPS core level spectra. Fig. 3(a1) shows the Cu 2p spectra, the peaks corresponding to the Cu $2p_{3/2}$ and Cu $2p_{1/2}$ are observed at around 932.5 and 952 eV respectively similar to Cu_2O spectra where Cu ions have a +1 valance state. Absence of any satellite line caused by multiplet splitting confirms that Cu is in $3d^{10}$ configuration [14-15].

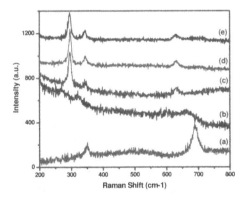

Figure 2. Temperature dependent Raman spectra of (a) single phase CFO at room temperature (300K), (b) CFO at 750 K, (c) disintegrated CFO during HT Raman measurement (at 300K), (d) disintegrated CFO during DTA (at 300K), and (e) disintegrated CFO during TGA (at 300K)

Raman scattering selection rules can be derived from the point group symmetry C_{3v} (space group $R\bar{3}m$). There are 12 optical phonon modes in the zone center (k ~ 0) classified as $\Gamma_{opt.R3c} = A_{1g}+E_g+3A_{2u}+3E_u$, out of which 2 phonons with A_{1g} and E_g symmetries are Raman active. $\Gamma_{opt.R3c} = A_{1g}+E_g$. A_{1g} modes represent the vibrations of Cu-O bonds along the c-axis whereas the doubly degenerate E_g modes describe the vibration along a-axis [9-10]. Figure 2 (a), (b), and (c) show the temperature dependent Raman spectra of single phase CFO at 300K, CFO at 750 K, and disintegrated CFO during HT Raman measurement (at 300K) respectively. Single phase CFO showed two Raman active modes at 349 cm^{-1} (E_g) and 690 cm^{-1} (A_{1g}) that matched with other delafossite structures. Raman spectra showed significant shift in both active modes towards lower frequency side and decrease in intensity with increase in temperature. Raman spectra of CFO showed almost similar modes as that of CFO below 750 K whereas above 750 K additional modes indicating disintegration of CFO at local level were observed. The thermally decomposed samples gave phonon modes at 294, 344, and 628cm^{-1} corresponding to (A_g), (B_{1g}), and (B_{2g}) symmetries of CuO [16-18] at room temperature.

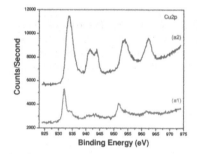 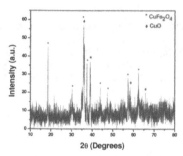

Figure 3. (a). X-ray photoelectron spectroscopy of pure CFO bulk (a1) and disintegrated sample (a2) for Cu ions and (b) XRD pattern of our disintegrated CFO sample. All of its major diffraction peaks are identified as either from $CuFe_2O_4$ or from CuO.

The XPS results on such samples are shown in Figure 3 (a2). The Cu $2p_{3/2}$ spectrum of disintegrated sample is wider than that of pure CFO. Also there is an intense satellite peak around 9eV above the main peak. This could be due to the multiplet splitting in the $3d^9$ electronic configuration. The Cu 2p spectrum is similar to that of CuO where Cu ions are divalent [14-15]. In comparison to the pure CFO, there is no change in the Fe 2p spectrum of the disintegrated sample. Hence we can conclude that Fe ions remain in the +3 state even after decomposition.

The XRD pattern of the disintegrated sample is shown in the Figure 3(b). It showed complete disintegration of CFO above 750 K, all the peaks matched well with either the CuO or $CuFe_2O_4$ sample's peaks [19-20]. XRD data of disintegrated CFO confirmed that above 750 K, in air CFO decomposes to form CuO and $CuFe_2O_4$ as per the following equation (1) [21].

$$2\,Cu^+Fe^{+++}\,O_{2+\delta} + (1-2\delta)/2\,O_2 \;\;=\;\; (Cu^{++}{}_{1/3}Fe^{+++}{}_{2/3})_{3+\varepsilon}\,O_4 + CuO \qquad (1)$$

TGA analysis of as synthesized CFO powder provides information about its thermal properties and helps to confirm the temperature at which thermal decomposition starts. Results are presented in Figure 4 (a). There were three main temperature regions of the mass change. The first one around 750K where the mass gain starts can be due to the absorption of O and commencement of disintegration. The second region, where the sample mass was almost constant i.e. around 1100K indicates almost completion of disintegration. Around 1200 K, sample started to lose its mass suggest further disintegration with another form. The DTA pattern

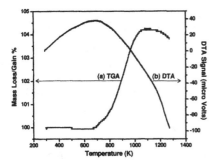

Figure4. TGA and DTA curves of as synthesized CFO samples

of CFO is shown in Figure 4 (b).The first region is exothermic and it extends up to 750K. Above this temperature the heat absorption and the disintegration process started. There is another change in slope in this endothermic region at 1200 K. Hence TGA and DTA analyses suggest that the CFO disintegration started above 750K and it completed around 1100 to 1150. Additionally, XPS, XRD and room temperature Raman (Figure 2 (d) and (e)) spectra were recorded for disintegrated samples from TGA and DTA analysis. The results were exactly similar to those obtained for HT Raman disintegrated sample.

CONCLUSIONS

Polycrystalline CFO ceramics were successfully synthesized by conventional solid state reaction technique. EDX and XPS indicate Cu and Fe are in +1 and +3 electronic state respectively. DTA and TGA suggests that disintegration of CFO started around 750 K but it reached in the mature phase at 1100 K. XRD and XPS of the disintegrated powder matched well with the XRD and XPS spectra of CuO and $CuFe_2O_4$ revealing CFO oxidation into cupric oxide and copper ferrite. Temperature dependent Raman spectra showed that CFO disintegrates around 750 K, matched with the Raman spectra of CuO. Raman spectra of all samples which went above 750 K during measurements showed only Raman spectra of CuO suggests and confirm commencement of disintegration of CFO around 750 K.

ACKNOWLEDGMENTS

This work is supported by the DEPSCoR project (Grant # W911NF-06-1-0183) and DoE project (Grant# DE-FG02-08ER46526).

REFERENCES

1. Hiroshi Kawazoe, Masahiro Yasukawa, Hiroyuki Hyodo, Masaaki Kurita, Hiroshi Yanagi, and Hideo Hosono, Nature, 389, 939 (1997).
2. Robert D. Shannon, Charles T. Prewitt, and Donald Burl Rogers, Inorg. Chem., 10 (4), 719 (1971).
3. M. C. Friedel, C. R. Hebd. Acad. Sci., 77, 211 (1873).
4. Kei Hayashi, Tomohiro Nozaki, and Tsuyashi Kajitani, Japanese Journal of Applied Physics, 46, 5226 (2007).
5. T. Kimura, J.C. Lashley, and A. P. Ramirez, Physical Review B, 73, 220401 (2006).
6. S. Seiki, Y. Yamasaki, Y. Shiomi, S. Iguchi, Y. Onose, and Y. Tokura, Physical Review B, 75, 100403 (2007).
7. H. Takahashi, Y. Motegi, R. Tsuchigane, and M. Hasegawa Journal of Magnetism and Magnetic Materials, 272, 216–217 (2004).
8. J.Pellicer-Porres, A. Segura, Ch. Ferrer-Roca, D. Martinez-Garcia, J.A. Sans and E. Martinez, J. P. Itié, A. Polian, F. Baudelet, A. Muñoz , P. Rodríguez-Hernández, and P. Munsch Physical Review B, 69, 024109 (2004).
9. J.Pellicer-Porres, A. Segura, E. Martinez, A. M. Saitta, A. Polian, J. C. Chervin, and B. Canny, Physical Review B, 72, 064301 (2005).
10. J.Pellicer-Porres, D. Martinez-Garcia, A. Segura, P. Rodriguez-Hernandez, A. Munoz, J.C. Chervin, N. Garro, and D. Kim, Physical Review B, 74, 184301 (2006).
11. K. Samanta, P. Bhattacharya and R.S. Katiyar, Physical Review B, 75, 035208 (2007)
12. "POWD" an interactive powder diffraction data interpretation and indexing program, Vr 2.1, E. Wu, school of Physical Science, Finder University of South Australia, Bedford, S 5042, Australia.
13. JCPDS pdf# 752146
14. J. Ghijsen, L. H. Tjeng, J. Van Elp, H. Eskes, J. Westerink, G. A. Sawatzky, and M.T. Czyzyk, Phys Rev B 38, 11322 (1988)
15. V.R. Galakhov, A.I. Poteryaev, E.Z. Kurmaev, V.I. Anisimov, St. Bartkowski, M. Neumann, Z. W. Lu, B. M. Klein and Tong-Rong Zhao Physical Review B, 56, 4584 (1997).
16. M.A. Dar, Q. Ahsanulhaq, Y.S. Kim, J.M. Sohn, W.B. Kima, H.S. Shin, Applied Surface Science 255, 6279–6284 (2009)
17. H.F. Goldstein, Dai-sik Kim, Peter Y. Yu, L.C. Bourne, J.P. Chaminade and Leon Naganga, Phys. Rev. B 41, 7192-7194 (1990)
18. M.H. Chou, S.B. Liu, C.Y. Huang, S.Y. Wu and C.L. Cheng, Applied Surface Science 254, 7539–7543 (2008)
19. JCPDS pdf# 801917
20. JCPDS pdf# 770010
21. E. Muginier, A. Barnabe and P. Tailhades, Solid State ionics, 177, 607-612 (2006).

Mater. Res. Soc. Symp. Proc. Vol. 1183 © 2009 Materials Research Society   1183-FF06-15

Effect of Partial Oxygen Pressure on Structural, Electrical, and Magneto Transport Properties of Cobalt Doped Indium Oxide Thin Films

A. Ghosh, N. Ukah, Ram Gupta, P. Kahol and K. Ghosh
Physics, Astronomy, and Materials Science, Missouri State University,
Springfield, Missouri

ABSTRACT

Dilute magnetic semiconductors are ferromagnetic semiconductors where the semiconductor host is doped with a small percentage of magnetic atoms. Recently it is reported that the structural and electrical properties of pure indium oxide can be modified by growth parameters. In this paper we investigate magneto-transport properties of Co-doped In_2O_3 dilute magnetic semiconductors thin films grown on sapphire and quartz substrates using pulsed laser deposition technique. The effect of partial oxygen pressure on structural, electrical, optical, and magneto-transport properties was discussed in details. The crystallinity of the films largely depends on growth temperature. Magneto-transport properties such as temperature dependent resistivity and magneto-resistance were found to be very sensitive to the micro-structural properties such as crystalinity as well as oxygen defect. The electrical carrier density of the films depends on oxygen pressure and a change of two orders of magnitude is observed. Depending on growth parameters, both positive and negative magneto-resistance is observed. Optical band-gap seems to vary with the growth partial oxygen pressure.

INTRODUCTION

Diluted magnetic semiconductors (DMS) are rare group of promising semiconductor compounds in which a fraction of the constituent ions is replaced by magnetic ions.[i] DMS offers a possibility of studying the magnetic phenomena in crystals with a simple band structure and excellent magneto-optical and transport properties. The magnetic properties of these semiconductors can be tuned not only by an external magnetic field but by varying the band structure and/or carrier, impurity and magnetic ion concentrations.[ii] It has been found that doping metal oxides such as ZnO, TiO_2, and In_2O_3 with magnetic ions such as Fe, Co, Mn, and Cr produces DMS, which exhibit ferromagnetism above room temperature.[iii] In_2O_3, a transparent opto-electronic material, is an interesting prospect for spintronics due to a unique combination of magnetic, electrical, and optical properties.[iv] Pure and doped indium oxide is a widely used transparent conducting oxide with tunable carrier concentration and mobility.[v,vi] Recently, high temperature ferromagnetism in manganese doped indium tin oxide grown using thermal evaporation is observed.[vii] Kim et al have also reported ferromagnetism in chromium doped indium oxide films using pulsed laser deposition technique.[viii] In this proceedings paper, we report the effect of growth temperature on structural, electrical, and magneto-transport properties of cobalt doped indium oxide films grown by pulsed laser deposition technique.

EXPERIMENTAL DETAILS

Cobalt doped indium oxide (ICO) target for producing thin films using pulsed laser deposition method is prepared by standard solid state reaction method. Required amount of In_2O_3 (99.99%) and Co_3O_4 (99.9%) were mixed to get the target having 5 at%. of cobalt. The powders were well mixed and sintered at 800 °C for 10 hours for three times. The mixture was brought at room temperature and pressed using a hydraulic press at 20 tons load, to get the circular disc shaped target of about 1cm radius . Thin films of ICO were grown on sapphire and quartz substrate using pulsed laser deposition technique (KrF excimer laser, λ=248 nm and pulsed duration of 20 ns). The laser was operated at a pulse rate of 10 Hz, with energy of 300 mJ. All the films were deposited under vacuum of base pressure 1×10^{-6} mbar. Thickness of the films grown in vacuum and in various partial oxygen pressures such as 1×10^{-4} mbar, 2.5×10^{-4} mbar, 4×10^{-4} mbar, 4.3×10^{-4} mbar, 1×10^{-3} mbar, to 1×10^{-2} are 130 nm, 127 nm, 122 nm, 116 nm, 112 nm, 90 nm, 70 nm, as measured by atomic force microscopy. The Crystallographic structures of the films were studied using Bruker AXS x-ray diffractometer (CuK$_\alpha$, λ =1.5406 Å). The Hall measurements were done using Van der Pauw technique as the magnetic field was varied from ±1.3 T. Electrical and magnetotransport characterizations were done using current source (Keithley 230) and voltmeter (Keithley 182). UV-Visible spectroscopic measurements have also been performed.

RESULTS AND DISCUSSION

The X-ray diffraction patterns of the ICO films grown on sapphire substrate at various growth partial oxygen pressures are shown in Figure 1. Lattice constant of ICO films was calculated using Bragg's diffraction law assuming simple cubic structure. Orientation of all the films along (222) direction and the phi-scan reveals that the films were highly crystalline. Calculated lattice constants for these films are 10.197 Å, 10.36 Å, 10.39 Å, 10.4 Å, and 10.4 Å, for films grown at 400°C in vacuum and at various other partial oxygen pressures such as 1×10^{-4} mbar, 2.5×10^{-4} mbar, 4×10^{-4} mbar, 4.3×10^{-4} mbar. Extra peak due to the doped cobalt into indium oxide was not observed, revealing the absence of any impurity phase in the films. Crystalline structure of sapphire is hexagonal (a = 4.758 Å), and that of indium oxide is cubic bixbyte (a = 10.197 Å). It is very interesting to see the kind of growth mechanism this system follows even with such a huge lattice mismatch. From XRD results we have seen that ICO grows epitaxially on sapphire (0001) substrate. Growth in this case is thought to be following domain matching epitaxy.

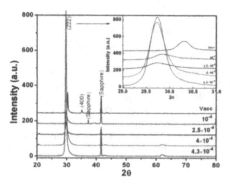

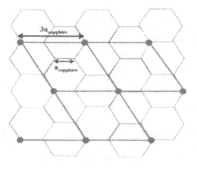

Figure 1. XRD patterns of ICO films grown on sapphire at different partial oxygen pressure at 400°C.

Figure 2. Schematic diagram, illustrating the mechanism of epitaxial growth of ICO thin films on sapphire substrate.

70

This mechanism involves matching of integral multiple (m) of lattice planes of the substrate with integral multiple (n) of a lattice plane (as shown in Figure 2) of the material in the film. Cubic bixbite is not a simple cubic structure but being very close to BCC structure. We have done all our calculations assuming a BCC structure for indium oxide with lattice constant a = 10.197 Å. Our calculation shows that the distance between the nearest neighbor atoms for indium oxide contained in (222) plane (BCC structure) is three times that of the lattice constant of sapphire. One can also proof that the angle between any three nearest neighbor atoms contained in (222) plane of indium oxide is 60°.

Figure 3 (a) illustrates the transmission spectra for the ICO films grown on quartz substrate at 400°C with various partial oxygen pressures. It is expected that the effect of partial oxygen pressure will remain same for ICO films grown on quartz and sapphire substrates. As the partial oxygen growth pressure increases the absorption edge shifts towards higher wavelengths. This can be observed clearly in Figure 3 (b), which shows a relationship between photon energy and absorption coefficient. The absorption coefficient α can be calculated from the relation $T = A \exp(-\alpha d)$, where T is the transmittance of thin film, A is a constant, and d is the film thickness. For the determination of optical band gap the Tauc model[ix] and Davis and Mott model[x] were applied. According to these models the following equation holds well in high absorption region: $\alpha h v = D(h v - E_g)^n$. Where, E_g is the optical band gap, $h v$ is the energy of the photon. Value of n is 1/2 or 2/3 for a direct transition.[xi,xii] For ICO thin films n=1/2 is suited most, since it gives the best linear curve in the band-edge region. Extrapolation of the linear region of the graph to the photon energy axis estimates the band gap E_g as shown in Figure 3 (b). With the increase of the partial oxygen growth pressure from vacuum to 1×10^{-2} mbar, the optical band gap red-shifted from 4.0 eV to 3.55 eV. Optical band gap values for vacuum and various partial oxygen pressures like 1×10^{-4} mbar, 2.5×10^{-4} mbar, 4×10^{-4} mbar, 1×10^{-3} mbar, to 1×10^{-2} mbar are found to be 4 eV, 3.82 eV, 3.69 eV, 3.625 eV, 3.62 eV, 3.55 eV, respectively. Although these samples were deposited on quartz, but the red-shift of the band gap should also hold for films deposited on sapphire.

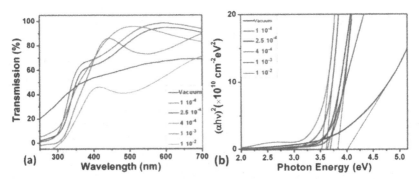

Figure 3. (a) Transmission spectra of ICO films grown on quartz substrate at 400°C, with different oxygen pressures and (b) Plot of $(\alpha h v)^2$ vs. photon energy for ICO films grown at 400°C with different oxygen pressures.

Measurement of the Hall coefficient is done via Hall effect measurement by using four probe method as shown in Figure 4. By measuring the thickness of the material we calculate the carrier density and figure out the type of carrier that we have in ICO films. The effect of partial oxygen pressure on electrical properties of the ICO thin films is discussed next. Figure 5 shows the dependence of electrical resistivity (ρ), carrier concentration (n), and mobility (μ) on partial oxygen pressure during film growth. The carrier concentration of the films was calculated from

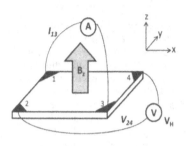

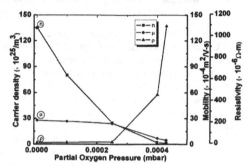

Figure 4. Schematic diagram of four probe van der Pauw method.

Figure 5. Effect of partial oxygen growth pressure on electrical resistivity (ρ), carrier concentration (n), and mobility (μ) of ICO films grown on sapphire substrate.

the Hall measurement, using the equation $n=1/eR_H$, where e is the electronic charge and R_H is the Hall coefficient[xiii], as the magnetic field was varied up to 1.3 T. It is observed that carrier concentration decreases with increase of partial oxygen pressure. Highest carrier concentration is achieved for the film grown in vacuum as $1.35 \times 10^{26}/m^3$ whereas lowest carrier concentration is found to be $1.25 \times 10^{24}/m^3$ for the film grown at 4.3×10^{-4} mbar. Resistivity of the films first increases with increasing partial oxygen growth pressure. Maximum mobility is found to be 27.8 $\times 10^{-4} m^2/V$-s for vacuum and the minimum is found to be 4.6 $\times 10^{-4} m^2/V$-s for the film grown under a partial oxygen pressure of 4.3×10^{-4} mbar.

Temperature dependent resistivity of ICO films grown at various partial oxygen pressures is shown in Figure 6 (a-d). As the temperature decreases the resistivity increases for all the ICO films. For the films which were grown at low partial oxygen pressure the increase in resistivity with decreasing temperature in not much. But for the films, which were grown at higher partial oxygen pressure the resistivity increases to a large amount with decreasing temperature. Magneto-resistive (MR) measurement reveals the presence of negative as well as positive MR in our ICO films at low temperatures (below 50 K). Film grown at 400°C at a partial oxygen pressure of 1×10^{-4} mbar shows negative MR (%$\Delta R/R$) with a max value of around -0.3%. For the film grown at 2.5×10^{-4} mbar we can see the presence of both positive as well as negative MR as shown in Figure 7 (a). Maximum positive MR is 0.8% which occurs at 5K with a maximum field of 1.3T. Magnitude of MR decreases as the temperature increases (at constant fields) and there is a transition from positive to negative to MR around 9K. Films grown under higher partial oxygen pressures shows large positive MR. Max positive MR of 8.9% is seen for the film grown

72

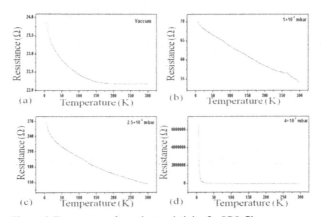

Figure 6. Temperature dependent resistivity for ICO films grown on sapphire at 400°C under various partial oxygen pressures.

at partial oxygen pressure of 4.3×10^{-4} mbar at 400C substrate growth temperature. Amount of negative MR is negligible in these films. But it is very much evident that there is a transition from positive MR to negative MR as the temperature increases. The deviation from the parabolic behavior at high fields and low temperatures is attributed to the conventional path-length-related positive MR, characteristic of traditional semiconductor transport.

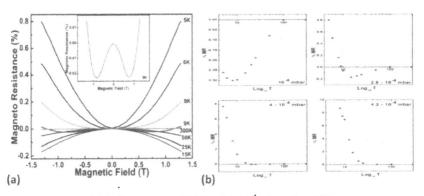

Figure. 7 (a) %MR of ICO film grown at 400°C at 2.5×10^{-4} mbar of partial oxygen pressure. (b) %MR vs $Log_{10}T$ plot for ICO films grown at various partial oxygen pressures.

The negative MR is not so unusual and has also been observed in other DMSs as well.[xiv] Mechanism involving this type of MR is proposed by Sivan, Entin-Wohlman, and Imry,[xv] which holds good for nonmagnetic semiconductors. This mechanism is based on the influence of magnetic field on quantum interference between many paths linking two hopping sites,[xvi] which obey the expression $\Delta R/R = T^{3/2}B^2$, at a small applied magnetic field in the variable range hoping regime. This is very much expected in presence of a coulomb gap in the density of states.[xvii] Another transport model helps explaining this negative MR, which is developed for magnetic semiconductors. This model considers the local exchange interaction between the magnetization and charge carriers and the fluctuation caused by it. The interaction introduces an additional energy disorder between different carrier hopping sites.

CONCLUSION

A series of cobalt doped (5%) indium oxide films were prepared using pulsed laser deposition technique. The effect of partial oxygen pressure on structural, electrical, and magnetic properties is studied in details. Hall Effect and temperature dependence resistivity measurements reveal that the electrical transport properties (carrier density, mobility, and resistivity) in these films can be tuned by varying partial oxygen pressure. U-V visible spectroscopy measurements reveal that the band gap of ICO films can also be tuned by varying the growth partial oxygen pressure. Magneto resistive measurement reveals the presence of positive as well as negative MR at low temperatures (below 50K). There is a transition from positive to negative to MR max positive MR of 8.9% is seen for the film grown at partial oxygen pressure of 4.3×10^{-4} mbar at 400C substrate growth temperature by applying a field of 1.3T. Magneto-transport data fit well with theoretical models of charge transport in nonmagnetic semiconductors.

REFERENCES

[i] P. K. Kahol, R. K. Gupta, K. Ghosh, Conference Proceedings, American Institute of Physics, 1063 (2008) 177.
[ii] I.B. Shim, C.S. Kim, J. Mag. Mag. Mater. 272-276 (2004) e1571.
[iii] G. Peleckis, X.L. Wang, S.K. Dou, J. Mag. Mag. Mater. 301 (2006)308.
[iv] T. Minami, MRS Bull. 25 (2000) 38.
[v] R. K. Gupta, K. Ghosh, S. R. Mishra, P. K. Kahol, J Optoelectron. Adv. Mater. 9 (2007) 2211.
[vi] C. Coutal, A. Azema, J.C. Roustan, Thin Solid Films 288 (1996) 248.
[vii] J. Philip, N. Theodoropoulou, J.S. Moodera, B. Satpati, Appl. Phys. Lett. 85 (2004)777.
[viii] H.S. Kim, S.H. Ji, S.K. Hong, D. Kim, Y.E. Ihm, W.K. Choo, Solid State Commun. 137 (2006).
[ix] J. Tauc, (Amorphous and Liquid Semiconductors), (Plenum, London, 1974).
[x] E. A. David, and N. F. Mott, Philos. Mag. 22, 903 (1970).
[xi] J. G. Lu, Z. Z. Ye, L. Wang, J. Y. Huang, and B. H. Zhao, Mater. Sci. Semicond. Process 5, 491 (2003).
[xii] S.T. Tan, X.W. Sun, X.H. Zhang, B.J. Chen, S.J. Chua, Anna Yong, Z.L. Dong, and X. Hu J. Cryst. Growth 290, 518 (2006).
[xiii] 10 L.J.Van der Pauw, Philips Res. Rep. 13 (1958) 1.
[xiv] A. Oiwa, A. Endo, Y. Iye, H. Ohno, and H. Munekata, Phys. Rev. B 59, 5826 (1999).
[xv] U. Sivan, O. Entin-Wohlman, and Y. Imry, Phys. Rev. Lett. 60, 1566 (1988).
[xvi] Y. Zhang and M. P. Sarachik, Phys. Rev. B 43, 7212 (1991).
[xvii] A. L. Efros, and B. I. Shklovskii, J. Phys. C 8, L49 (1975).

MTJ and Dynamics

Mater. Res. Soc. Symp. Proc. Vol. 1183 © 2009 Materials Research Society   1183-FF07-02

Tailoring TMR ratios by ultrathin magnetic interlayers: A first-principles investigation of Fe/MgO/Fe

Peter Bose,[1,3] Peter Zahn,[1] Jürgen Henk,[2] and Ingrid Mertig[1,2]

[1]Physics Institute of the Martin Luther University Halle-Wittenberg, Theory Department, von Seckendorff-Platz 1, D-06120 Halle (Saale), Germany
[2]Max Planck Institute of Microstructure Physics, Theory Department, Weinberg 2, D-06120 Halle (Saale), Germany
[3]International Max Planck Research School for Science and Technology of Nanostructures, MPI of Microstructure Physics, Weinberg 2, D-06120 Halle (Saale), Germany

ABSTRACT

For spintronic device applications, large and in particular tunable tunnel magnetoresistance (TMR) ratios are inevitable. Fully crystalline and epitaxially grown Fe/MgO/Fe magnetic tunnel junctions (MTJs) are well suited for this purpose and, thus, are being intensively studied [1]. However, due to imperfect interfaces it is difficult to obtain sufficiently large TMR ratios that fulfill industrial demands (e.g. [2]).

A new means to increase TMR ratios is the insertion of ultra-thin metallic buffer layers at one or at both of the Fe/MgO interfaces. With regard to their magnetic and electronic properties as well as their small lattice mismatch to Fe(001), Co and Cr spacer are being preferably investigated.

We report on a systematic first-principles study of the effect of Co and Cr buffers (with thicknesses up to 6 ML) in Fe/MgO/Fe magnetic tunnel junctions (MTJs) on the spin-dependent conductance. The results of the transport calculations reveal options to specifically tune the TMR ratio. Symmetric junctions, i.e. with Co buffers at both interfaces, exhibit for some thicknesses much larger TMR ratios in comparison to those obtained for Fe-only electrodes. Further, antiferromagnetic Cr films at a single interface introduce TMR oscillations with a period of 2 ML, a feature which provides another degree of freedom in device applications. The comparison of our results with experimental findings shows agreement and highlights the importance of interfaces for the TMR effect.

INTRODUCTION

Fully crystalline Fe/MgO/Fe MTJs show very high TMR ratios [3-6]. After intensive studies of these systems, the research was gradually extended to other promising systems. MgO tunnel junctions with amorphous CoFeB electrodes for instance were found to improve structural and magnetic properties, resulting in giant TMR ratios [1].

The detailed structure of the interfaces in Fe/MgO/Fe essentially determines the spin-polarized current. Thus, it is obvious to manipulate the interfaces in a controlled way to achieve larger TMR ratios. Considering the magnetic profiles in Figs. 1 and 2, it is expected that Cr and Co buffers have a sizable effect on the tunnel current, and especially on its spin-polarization.

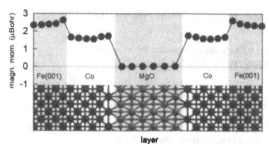

Figure 1. Layer-resolved magnetic moments of bcc Fe(001)/x(Co)/6MgO/x(Co)/Fe(001) magnetic tunnel junction with $x = 6$. The geometry of the MTJ is sketched at the bottom.

Magnetic tunnel junctions with bcc Co electrodes are theoretically predicted [7] to provide much larger TMR ratios than those with Fe electrodes. However, Co grows only up to a few monolayers on MgO in the bcc phase; for thicker layers a structural transition to the hcp structure takes place, thus introducing imperfections which definitely reduce the TMR ratio. One aim of this work is to investigate whether thin Co interlayer in Fe/MgO/Fe increase the TMR ratios, similar to those ratios predicted with infinite Co leads and comparable with those obtained with CoFeB electrodes.

Nagahama et al. [8] showed that the insertion of Cr films in MTJs with amorphous AlO tunnel barriers and Fe leads exhibits a 2-ML oscillation of the experimental TMR ratio as a function of Cr thickness. A 2-ML oscillation is a signature of a layerwise antiferromagnetic order in the Cr film, in agreement with theoretical findings for Mn buffers [9]. In the latter work, the even-odd effect in the sign of the TMR ratio was attributed to the atomic Mn layer adjacent to the tunnel barrier: its magnetization direction plays a key role in the spin-dependent electronic transport.

Cr couples antiferromagnetically to Fe(001) and shows layerwise AFM order (Fig. 2). In addition, a large magnetic moment is found at the interface with MgO. This finding raises the question whether a single Cr spacer at a single interface produces a defined sign reversal of the TMR ratio.

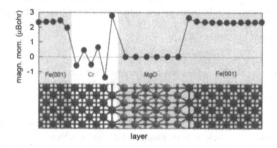

Figure 2. As Figure 1, but for bcc Fe(001)/x(Cr)/6MgO/ Fe(001) MTJ with $x = 6$.

78

THEORETICAL BACKGROUND

In a first step, *ab-initio* electronic-structure calculations were performed within the framework of the local spin-density approximation to density functional theory. The so achieved self-consistent potentials serve as input for the subsequent transport calculations. Both electronic-structure and transport properties are obtained by a Korringa-Kohn-Rostoker multiple-scattering Green's function formalism. Due to its perfect adaptation to the planar geometry a layer-KKR computer code was used for the electronic transport calculations. The self-consistent treatment of the Fe(001)/x(Co)/MgO/x(Co) /Fe(001) and Fe(001)/ x(Cr)/MgO/Fe(001) MTJs, x = 1, ..., 7 ML, follows closely those for Fe(001)/MgO/Fe(001) reported in [10]. In particular, atomic positions and interlayer distances were taken from experiment [11]. So, slight changes are expected due to the different atomic volumes of Co and Cr with respect to Fe. The number of MgO layers was fixed for each set-up to 6 ML (corresponding to a thickness of 10.7 Å).

Within the Landauer-Büttiker approach [12], the zero-bias conductance is calculated in terms of the transmittances $T(E)$ at the Fermi level. The latter is computed by integrating the wavevector-resolved transmittances $T(E, k_\parallel)$ over the two-dimensional Brillouin zone (2BZ) [13], where $T(E, k_\parallel)$ is the sum of the transmission probabilities of all Bloch states in the leads. Since both setups exhibit *4mm* symmetry, the number of wavevectors $k_\parallel$ in the 2BZ integration was reduced from 80 000 equidistant mesh points to about 10000 of the irreducible part while maintaining the same level of accuracy.

The TMR ratio is expressed by the asymmetry of the conductances for the parallel (G_P) and antiparallel (G_{AP}) magnetic configurations of the Fe electrodes, normalized by the conductance of the AP case ('optimistic TMR ratio').

RESULTS AND DISCUSSION

Co interlayers at both Fe/MgO interfaces

The conductance for the parallel configuration G_P is almost constant with an apparent 2-ML oscillation, with maximum (minimum) conductance for an even (odd) number of x ML. In contrast G_{AP} shows a more complex thickness dependence (Fig. 3). G_{AP} starts approximately two orders of magnitudes smaller than GP at x = 0 ML, but reaches a pronounced maximum an order of magnitude larger at a thickness of two Co layers. For x = 3 – 5 ML it decreases and reaches nearly the level obtained without Co spacers. Another, but some smaller maximum is obtained for one additional Co layer (x = 6 ML). In comparison to the Fe/MgO/Fe MTJ without Co film, a sizably smaller G_P but larger G_{AP} value is achieved for infinite Co electrodes.

The calculated TMR ratios exhibit three noticeable characteristics. Firstly, 3 and 5 ML thick Co interlayers lead to huge TMR ratios: 10000% at 3 ML and 15700% at 5 ML – which are significantly larger than the 6800% obtained without Co spacers (indicated by the green horizontal line in Fig. 3). Secondly, the TMR at 2 ML Co drops as a consequence of the large G_{AP} value, caused by interface resonances, nearly to zero. Thirdly, a much smaller TMR value is calculated for infinite Co leads in comparison to that determined with pure Fe leads, a finding in contrast to results reported in [7]. This may be related to differences in the geometries. Previous investigations of Fe/MgO/Fe systems have shown that slightly differing atomic positions in the interface region can lead to sizably different conductances and TMR ratios.

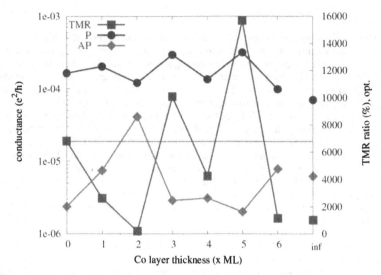

Figure 3. Conductances for the P (black circles) and AP (blue diamonds) magnetic configurations of Fe/x(Co)/MgO/x(Co)/Fe versus Co thickness. The 'optimistic' TMR ratio is shown as red squares. Results for Co electrodes, replacing the Fe electrodes are shown in addition ('inf').

Cr interlayers at one Fe/MgO interface

Fig 4a. displays in analog to Fig 3., the thickness dependence of the P and AP conductance with up to 7 Cr layers. Both, G_P and G_{AP} exhibit an exponential decay as a function of the Cr thickness x. The G_P decay rate is hereby visibly larger than that one for G_{AP}.

G_P and G_{AP} reveal superimposed to the exponential decay, even-odd oscillations that are in antiphase. These characteristics can be traced back to the layer-wise antiferromagnetism of the Cr layers. In Fig 2. the exemplary case with $x = 6$ ML shows that Cr couples layer-wise antiferromagnetically to the Fe(001) substrate. It turns out and can be seen too in Fig 2. that the magnetic layer at the Cr/MgO interface possesses the largest uncompensated local magnetic moment of the Cr spacer. It can be deduced from previous tunnel magneto-resistance investigations with layer-wise antiferromagnetic Mn films [9] that the Cr film acts generally as a spin-filter for the electron currents. But the decisive influence which changes the spin-polarization of the currents can be directly addressed to the magnetic atom adjacent to the MgO barrier.

With help of Fig 4b, which displays the local magnetizations of these interface layers, it is clear that an even number of Cr layers leads to a positive local magnetic moment which results directly in local maxima (minima) for G_P (G_{AP}). Vice versa, negative local moments for odd Cr layers cause a local maxima (minima) for G_{AP} (G_P).

The periodic maxima and minima of G_P and G_{AP} cause a pronounced even-odd effect with periodic changes of $G_P > G_{AP}$ and $G_P < G_{AP}$. Consequently, this results in an oscillation of the TMR ratio shown in Fig 4a. This oscillation with a period of 2-ML is connected to a periodic sign reversal of the TMR ratio.

The TMR ratio with no Cr spacer is about 6800%. This order of magnitude shows up again only for a Cr thickness of 2 monolayers. In particular we would like to emphasize that the TMR value of approximately 8200% is larger than that one found for Fe/MgO/Fe MTJs without any Cr interlayers. The TMR ratio for 1 ML is strongly reduced about two orders of magnitude. Apart from the large amplitude for 2 ML, this reduced level is reached and maintained – alternating between about plus-or-minus 100% – for all thicker Cr films.

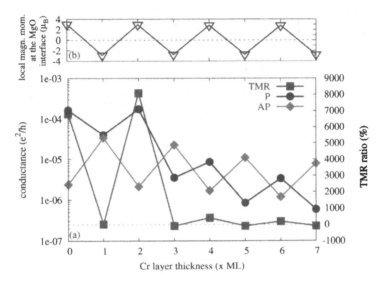

Figure 4. (a) As Figure 3, but for Fe/x(Cr)/MgO/Fe versus Cr thickness. (b) Local magnetic moment of the magnetic layer at the interface for each Cr thickness x.

CONCLUSIONS

Bcc Co interlayers at both interfaces of Fe/MgO/Fe junctions do not *per se* improve TMR ratios as compared to Fe/MgO/Fe junctions. Only specific Co thicknesses, namely 3 and 5 ML, result in larger TMR ratios.

The insertion of a single, layer-wise antiferromagnetic Cr buffer causes 2-ML oscillations of the conductances as a function of the Cr thickness, which show up as an even-odd change of the TMR's sign. The TMR ratio is generally, compared to the case with no Cr spacer, reduced about two orders of magnitude. Only a 2 ML thick Cr spacer is found to reproduce a sizeable larger TMR ratio.

ACKNOWLEDGEMENTS

One of us (PB) acknowledges support by the International Max Planck Research School for Science and Technology of Nanostructures. We thank R. Matsumoto and S. Yuasa (AIST, Tsukuba, Japan) for fruitful discussions.

REFERENCES

[1] S. Yuasa, and D. D. Djayaprawira, J. Phys. D: Appl. Phys. **40** (2007), R337-R354
[2] P. Bose, A. Ernst, I. Mertig, and J. Henk, Phys. Rev. B **78**, 092403 (2008)
[3] W. H. Butler, X.-G. Zhang,T. C. Schulthess and J. M. MacLaren, Phys. Rev. B **63**, 054416 (2001)
[4] J. Mathon and A. Umerski, Phys. Rev. B **63**, 220403 (2001)
[5] S. Yuasa, T. Nagahama, A. Fukushima, Y. Suzuki, and K. Ando, Nat. Mat. **3**, 868 (2004)
[6] Parkin et al., Nat. Mat. **3**, 862 (2004)
[7] X.-G. Zhang and W. H. Butler, Phys. Rev. B **70**, 172407, (2004)
[8] T. Nagahama, S. Yuasa, E. Tamura, and Y. Suzuki, Phys. Rev. Lett. **95**, 086602, (2005)
[9] P. Bose, I. Mertig, and J. Henk, Phys. Rev. B **75**, 100402(R) (2007)
[10] Ch. Heiliger, P. Zahn, B. Yu. Yavorsky and I. Mertig, Phys. Rev. B **73**, 214441 (2006)
[11] Ch. Tusche et al., Phys. Rev. Lett. **96**, 119602 (2006)
[12] Y. Imry and R. Landauer, Rev. Mod. Phys. **71**, S306 (1999)
[13] J. M. MacLaren, X.-G. Zhang, W. H. Butler and X. Wang Phys. Rev. B **59**, 5470 (1999)

Poster Session: Spintronics Transport and Imaging

Mater. Res. Soc. Symp. Proc. Vol. 1183 © 2009 Materials Research Society   1183-FF10-12

Morphologic, compositional and magnetic characterization of sputtered CoCr thin films for applications in MTJs as hard spin injectors

P. Martino[1], A. Chiolerio[1], P. Pandolfi[2], P. Tiberto[3], P. Allia[2]

[1]Physics Department, Politecnico di Torino, Corso Duca degli Abruzzi 24,
IT-10129 Torino, Italy
[2]Materials Science and Chemical Engineering Department, Politecnico di Torino, Corso Duca degli Abruzzi 24,
IT-10129 Torino, Italy
[3]Electromagnetism Division, INRIM, Strada delle Cacce 91,
IT-10135 Torino, Italy

ABSTRACT

Spin-polarized currents across an insulating tunnel barrier, needed for the development of efficient magnetic tunneling junctions (MTJs), may be obtained using hard spin injector electrodes. Thin films of CoCr solid solutions have been fabricated involving two main steps: 1) deposition of Co/Cr alternated layers via RF magnetron sputtering both onto silicon (100) substrates and thermally oxidized wafers and 2) thermal annealing in a partial Ar pressure of 1 mTorr at 450°C for 1 hour and cooling treatment in a uniform magnetic field (600 Oe). The deposition of stacks of pure elements and subsequent diffusion treatment has been preferred instead of the direct deposition of the native alloy because in the former case the right composition and magnetic bias may be tuned more easily playing on the layer thicknesses and number of repetitions.

A detailed numerical correlation of field effect SEM images and EDX micro-maps was used to evaluate the oxygen diffusion on the magnetic film, while an alternating gradient force magnetometer (AGFM) allowed us to evaluate at room temperature both coercivity and magnetic bias obtained after the field cooling treatment. The effect of standard thermal treatment on the homogeneity of the films is discussed, and a possible alternative heating technique is proposed.

INTRODUCTION

In magnetic tunneling junctions the spin injector electrode is usually a hard ferromagnetic (FM) layer exchange biased by an antiferromagnetic (AFM) layer [1].
It has been demonstrated that a phase separation occurs in CoCr alloy thin films [2] where the Co-rich regions dispersed in a Cr antiferromagnetic matrix seem to be the main reason for the increase of coercivity.
In this paper we discuss about the possible use of a thin film of CoCr multilayer transformed into solid solution as an effective alternative to a CoCr alloy system used as a hard spin injector in spintronics devices. Since the exchange bias is essentially an interfacial phenomenon occurring at the interface between Co and Cr grains, the magnetic properties may be easily tuned playing on each layer thickness and number of repetitions.
We report on a morphological and magnetic analysis of two different types of sputtered Co/Cr multilayer samples comparing them with a CoCr solid solution, obtained after annealing and subsequent field cooling, in order to draw information about the mechanism of metal inter-

diffusion, the magnetic anisotropy with the final aim of realizing effective spin-polarizing layers for MTJs.

EXPERIMENT

Two different Co-Cr stacks have been deposited by RF magnetron sputtering at a pressure of 10 mTorr of Ar, fixing the Co to Cr ratio (5:1) and varying the total thickness (45 and 70 nm). Thermally oxidized substrates have been used; the residual anisotropy given by free growth on the amorphous silica has been measured. 1 cm^2 wide multilayers have been realized, consisting of the following multilayer sequences: $(Co_{10}/Cr_2)^3Co_9$ (referred to as C1) and $(Co_{2.9}/Cr_{0.6})^{20}$ (referred to as C2), where the subscript represents the thickness in nm and the superscript the number of repetitions. The thermal treatment was done in a partial Ar pressure of 1 mTorr (base vacuum ~10^{-7} Torr) to allow heat diffusion at 450°C for 1 hour on sample C1 (the annealed C1 sample is referred to as C3).
The cooling of the samples was done in a uniform magnetic field of 600 Oe generated by an electromagnet.
Morphology and grain texture of samples were investigated by Field Effect Scanning Electron Microscopy (FESEM) and a detailed numerical correlation of FESEM images and Energy Dispersive X-ray (EDX) micro-maps were used to evaluate the oxygen diffusion on the solid solution after annealing and the metal distribution within the annealed sample.
Alternating Gradient Force Magnetometer (AGFM) measurements performed at room temperature up to 5 kOe allowed us to evaluate both coercivity and magnetic bias obtained after the field cooling treatment of our samples.

DISCUSSION

The RF sputtering deposition has been experimented, together with a reduction of the annealing time, to avoid poor surface homogeneity with respect to other techniques, such as the thermal evaporation; the thickness of each layer has been selected to promote interlayer diffusion after the thermal annealing.

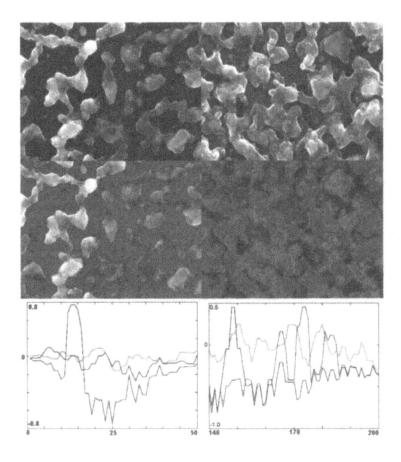

Figure 1. A and B: FESEM images of sample C3 surface, on different sites. The area represented in panels A and B has been also used to get EDX spectra with 2-dimensional resolution. C and D: color-hatched areas have been chosen for the correlation analysis. E and F: output of the correlation, spectra of oxygen (red), Chromium (green) and cobalt (blue), after subtraction of the silicon spectrum background; on the vertical axis the cross-correlation factor (1.0 for completely correlated, -1.0 for completely uncorrelated), on the horizontal axis the selected area gray-scale value.

A detailed numerical correlation of FESEM images and EDX micro-maps has been used to evaluate the oxygen diffusion from the substrate towards the annealed magnetic film C3, in order to exclude the formation of mixed magnetic oxides. The surface appears constituted by separated islands (figure 1 A and B), due to de-wetting from the thermally oxidized silicon substrate. Depending on the gray-scale value, FESEM-EDX micro-map correlations have been computed making use of a technique described in detail in [3]. Both panel C and D show that a color transparent-hatched region has been chosen to compare EDX spectra of O, Co and Cr: from 0 to 50 and from 155 to 255 gray threshold[1], respectively, corresponding to regions that are in the middle of CoCr islands, and regions that are on top of the islands; as it may be seen in panels E and F, the oxygen peaks are not over-imposed on the Co or Cr peaks, demonstrating that no diffusion occurred. Furthermore, it is possible to see that also the Co and Cr peaks, analyzed in this greyscale-evaluated elemental spectrum, are not fully over-imposed.

This means that at the nanoscale, regions with different compositions are found. It has not been possible to give FESEM / EDX numerical correlation outputs, since when the sample plane is observed from the top, it appears completely homogeneous and the Co and Cr signals feature a uniformly diffused statistics; on the contrary, when observing the samples along a fracture plane in the cross-section arrangement, the thickness of each layer being smaller than 10 nm, the electron beam generated by the field-effect gun is not sharp enough to allow collecting a significant statistics.

An AGFM apparatus has allowed to evaluate at room temperature both coercivity and magnetic bias obtained after the field cooling treatment (figure 2): referring to the geometrical easy axis of the samples, the annealing resulted in a reduction of coercivity from 52 to 15 Oe, with a great increase of the bias field, from 1.5 to 10 Oe, producing a visible shift of the hysteresis loop towards positive fields. This bias may be due also to the exchange coupling provided by AFM Cr domains in close contact with ferromagnetic CoCr grains or pure Co grains.

---

1   both FESEM and EDX images have been converted to 16 bit maps, hence featuring 256 grey values.

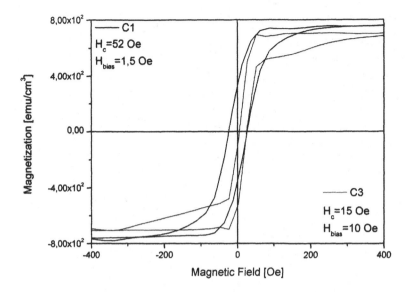

Figure 2. AGFM magnetization measurements of sample C1 (black line) and C3 (red line). Indicated in the graph frame the values of coercivities and magnetic bias.

For what concerns sample C2, the particular choice of layers thicknesses resulted in a well different magnetic behavior, as shown in figure 3 by a dataset composed of in plane and out of plane measurements (field applied orthogonally with respect to the substrate plane). In this case the sample is extremely soft in the plane of the film, with a coercive field around 250 mOe, and shows a slowly saturating response in the out of plane analysis, where the magnetization process is concluded at fields much higher than the explored region (figure 4).

89

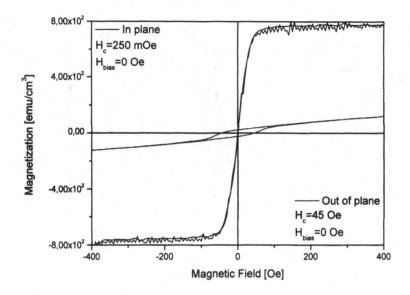

Figure 3. AGFM magnetization measurements of sample C2. Indicated in the graph frame the values of coercivities for both the in plane measure (black line) and out of plane one (red line).

Comparing figure 2 and figure 3, it is possible to notice that no magnetic bias (sample C2) or very little (sample C1) are found for the as deposited samples, while a net bias develops after diffusion annealing (sample C3). The absence of a detectable bias is possibly a consequence of a structural disorder found in the lattices of those Co layers which have been deposited on top of (111) Cr layers. In the solid solution produced by thermal annealing, the interface-to-bulk volumic ratio is much higher than in the original multilayers; furthermore, after the formation of the island-like zones (figure 1), a significant increase of the coercivity seems justified. A very recent work has shown the importance of the morphological anisotropy in ferromagnetic / antiferromagnetic exchange coupled systems [4].

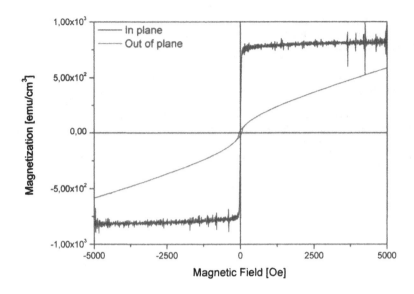

Figure 4. Full measure extended up to 5 kOe, for sample C2.

CONCLUSIONS

Morphological and magnetic properties of CoCr multilayers and solid solutions were investigated to compare them with CoCr alloyed systems proposed as single hard magnetic layer in ferromagnetic tunneling junctions [2].

The deposition of stacks of pure elements and subsequent diffusion treatment has been preferred instead of the direct deposition of the native alloy because in the former case the magnetic properties (coercivity and bias) may be easily tuned playing on the layer thicknesses and number of repetitions.

We demonstrated that the annealing treatment causes the metal interdiffusion and generates a final structure with two different Co and Cr phases similar to compositional regions that are responsible, in CoCr alloyed thin films, for the increase in coercive field. We found a different magnetic behavior for pre-annealed samples, showing a coercivity bigger than the annealed and field cooled one. Choosing thinner layers and increasing the repetition number, we found an extremely soft behavior in the pre-annealed condition.

Further experiments for the annealing and cooling conditions of extremely soft samples are therefore in progress.

In particular, we foresee the possibility of employing a laser equipment in order to induce a flash thermal treatment and obtain a much more homogeneous surface together with the correct metal interdiffusion and induced magnetic hardness.

REFERENCES

[1] J.S. Moodera and G. Mathon, *J. Magn. Magn. Mater.* 200 (1999) 248;
[2] A. Pundt, C. Michaelsen, *Phys.Rev.B.* 56, 22, (1997) 14352;
[3] A. Chiolerio, A. Chiodoni and P. Allia, Thin Solid Films, 516 (2008) 8453;
[4] F. Bisio, L. Anghinolfi, M. Canepa and L. Mattera, *Phys.Rev.B* 79 (2009) 054407.

Mater. Res. Soc. Symp. Proc. Vol. 1183 © 2009 Materials Research Society   1183-FF10-14

Electronic Structure and Magnetic Properties of Ti-Doped ZnO

S. Lardjane, G. Merad and H. I. Faraoun
Laboratory of Materials Discovery, Unit of Research Materials and Renewable Energies,
A. Belkaid University of Tlemcen, B.P. 119, 13000, Algeria

ABSTRACT

Recent experiments suggest that Ti doped ZnO can be a promising room temperature dilute magnetic semiconductor (DMS) and a potentially useful material for spintronic devices. Furthermore, the fact that Ti doped ZnO shows ferromagnetic behaviour despite it contains no magnetic element makes this system good candidate for theoretical investigation regarding the controversies about the origin of ferromagnetic ordering in TM-doped ZnO. In this work, the density functional theory (DFT) is used to calculate the electronic and magnetic structures of Ti-doped ZnO. The obtained results are used to discuss the origin of the ferromagnetism, and the contribution of different atoms in the magnetic moment.

INTRODUCTION

In recent years, the manipulation of spin in semiconductors (so-called spintronics) presents a new paradigm for functionality in electronic materials. As a semiconducting material, ZnO offers significant potential in providing charge, photonic, and spin based functionality. ZnO is a direct wide band gap semiconductor with potential utility in UV photonics and transparent electronics. ZnO doped with transition metal ions has been considered recently as a dilute magnetic semiconductor (DMS) [1]. While some ones claimed that transition metal-doped ZnO was ferromagnetically ordered, others suggested that inclusions of ferromagnetic compounds were responsible for the observed ferromagnetism, and that the semiconductor exhibited no ferromagnetism at all. Theoretically, unfortunately, ferromagnetism in semiconductors is rare and poorly understood, with ferromagnetic transition temperatures well below room temperature for most known materials. In particular, for Ti-doped ZnO there exist some inconsistent results; Imai and Watanabe [2] argued that there is no magnetic order in Ti-doped ZnO, while the recent experimental findings of Venkatesan et al. [3] and Rao et al. [4] showed the ferromagnetism in Ti-doped ZnO.
Clearly, for a successful synthesis of high T_C ferromagnetic DMSs, it is necessary to understand the origin of the ferromagnetism in these materials. More generally, the study of any semiconducting material that supports spin polarized electron distributions would be useful in understanding and developing spintronic concepts. In the present paper, we tried to clarify the ferromagnetism of Ti-doped ZnO and achieve sufficient knowledge of its properties using first-principles spin polarized calculations.

COMPUTATIONAL METHOD

Calculations presented here were carried out using the VASP pseudo potential program [5]. The pseudo potential technique is based on density functional theory (DFT) [6], in which theory the total ground state energy of a material is expressed as a unique functional of the electron density plus ionic contributions. In practice, a set of single-particle wavefunctions [7] are used to construct the electronic density; usually and for calculation convenience these are expressed in a plane-wave basis, where the plane-wave states are chosen to have the same periodicity as the simulation cell. This basis set is virtually infinite, but the coefficients associated with the high-energy basis states are negligible so cut-off energy is defined and only plane-wave states with lower energies are accepted into the basis. As the core electrons are relatively inert because strongly bounded, the pseudo-potential method treats these as to be frozen and concentrates the computational effort on the valence electrons only.

In principle, the total energy of the ground state is a functional of the density, although this functional is not known; this leads to the use of a series of self consistent calculation. While most of the energy contributions can be calculated exactly, the exchange-correlation energy cannot, and so we must use an approximation. In the calculations presented here the generalized gradient approximation of Perdew et al. [8] has been used to describe this exchange and correlation within the system.

In this work, Monkhorst-Pack [9] grid gamma centred was used to sample the Brillouin zone. Total energy convergence has been examined with respect to the number of k points to ensure a sufficiently high density of sampling point particularly required for metals. 4x4x4 grid was found to be sufficient.

ZnO has a wurtzite structure in which anions and cations, respectively, form two hexagonal close packed lattices separated from each other along the c-axis by the internal parameter u. In order to simulate the $Zn_{1-x}Ti_xO$, the supercell approach was employed. We thus extended the four atoms original ZnO unit cell to a 2*2*2 supercell containing 32 atoms (16 ZnO molecules). Within the super cell, one Zn atom is substituted by a Ti one. This leads to a doping level of x=0.0625.

RESULTS AND DISCUSSION

The total magnetic moment for the $Zn_{0.9375}Ti_{0.0625}O$ super cell is found to be 0.91 μ_B. The contribution of the Ti atom in the magnetic moment is of 0.884 μ_B, while the oxygen of the tetrahedron base, which are the nearest neighbours of the titanium one, make insignificant contributions of 0.003 μ_B, 0.003 μ_B and 0.002 μ_B ; the atom of the tetrahedron top makes the negative contribution of -0.028 μ_B. Those results are summarized in table I.

Manifestly, the Ti dopant is found to be the main contributor to the ferromagnetism of the doped ZnO. We point out that this is a very interesting result for there is no magnetic ion in this system. This suggests that the ferromagnetism originates in the exchange interaction between host and Ti atoms.

Table I The magnetic moments of the supercell (M_{total}), the Ti atoms (M_{Ti}), and the O atoms (M_O) for $Zn_{0.9375}Ti_{0.0625}O$

	Magnetic moments	
M_{total} (μ_B)	0.91	
M_{Ti} (μ_B)	0.884	
M_O (μ_B)	3 O 0.003 0.003 0.002	1 O -0.028

In order to clarify the origin of ferromagnetism in TM –doped semiconductors, the double exchange mechanism has been frequently proposed [11, 12]. Accordingly, we calculated the total density of states (DOS) and partial densities of states (PDOS). Results are shown in Figure 1.

It is found that the conduction electrons at the Fermi level are almost 100% spin polarized, indicating a nearly half-metallic band structure. On the basis of the PDOS of titanium atom, it is found that states near Fermi energy are composed mostly of Ti 3d bands, hybridized strongly with the O 2p ones. The Ti 3d states are located in the bottom part of conduction band region, which is just the origin of impurity band in the Ti-doped ZnO.

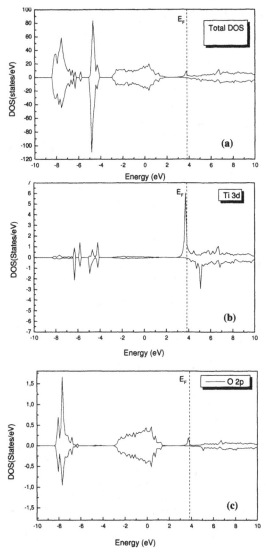

Figure 1. Total (a) and partial densities of states (b) Ti-3d and (c) O-2p of $Zn_{0.9375}Ti_{0.0625}O$. Positive (negative) values correspond to the majority (minority) spin

Due to the large amplitude of the d states at the Fermi level as can be seen on the figure, 3d electrons in the partially occupied 3d-orbitals of Ti are allowed to hop to 3d-orbitals of neighboring Ti, if this one has its magnetic moment oriented in the same direction. As a result, the d electron lowers its kinetic energy by hopping in a ferromagnetic order. Accordingly, we suggest that the ferromagnetic ground state observed in Ti-doped ZnO can be explained in terms of Zener's double exchange mechanism as has already been proposed by Sato et al. [14].

CONCLUSIONS

We carried out first-principles spin polarized calculations of the electronic and magnetic structures of Ti-doped ZnO. According to our results, ferromagnetic ordering occurs despite Ti is nonmagnetic in its natural phase. Furthermore, the analysis of total and partial densities of states allows us to suggest that the origin of the ferromagnetism in the Ti-doped ZnO can be understood by the double-exchange mechanism, controlled by the 3d Ti electrons that lower the total energy by hopping between adjacent levels in a ferromagnetic ordering.

The manifestation of ferromagnetism in this system suggests that some elements which are nonmagnetic in their natural phases can be chosen for fabrication of transparent ferromagnets.

REFERENCES

1. K. Ueda, H. Tabata, T. Kawai, Magnetic and electric properties of transition-metal-doped ZnO films, Appl. Phys. Lett. 79 (2001) 988–990.
2. Y. Imai, A. Watanabe, Comparison of electronic structures of doped ZnO by various impurity elements calculated by a first-principle pseudopotential method, J. Mater. Sci. Mater. Electron. 15 (2004) 743–749.
3. M. Venkatesan, C.B. Fitzgerald, J.G. Lunney, et al., Anisotropic ferromagnetism in substituted zinc oxide, Phys. Rev. Lett. 93 (2004) 177206–177209.
4. C.N.R. Rao, F.L. Deepak, Absence of ferromagnetism in Mn- and Co-doped ZnO, J. Mater. Chem. 15 (2005) 573–578.
5. G. Kresse and J. Hafner, Phys. Rev. B 47, 558 (1993); ibid. 49, 14 251 (1994). G. Kresse and J. Furthmüller, Comput. Mat. Sci. 6, 15 (1996). G. Kresse and J. Furthmüller, Phys. Rev. B 54, 11 169 (1996).
6. G. Kresse and J. Hafner, J. Phys.: Condens. Matt. 6, 8245 (1994).
7. P. Hohenberg and W. Kohn. Phys. Rev. B3 136 (1964) 864-871.
8. J. P. Perdew and Y. Wang, Phys. Rev. B45 (1992) 13244.
9. H.J. Monkhorst and J.D. Pack, Phys. Rev. B 13 (1976) 5188.
10. R.W.G.Wyckoff, Crystal Structures, Vol. 1, 2nd Edition, Wiley, New York, 1986 p.112.
11. K. Osuch, E.B. Lombardi, W.I. Gebick, First principles study of ferromagnetism in $Ti_{0.0625}Zn_{0.9375}O$, Phys. Rev. B 73 (075202) (2006) 1–5.
12. K. Sato, H. Katayama-Yoshida, Material design for transparent ferromagnets with ZnO-based magnetic semiconductors, Jpn. J. Appl. Phys. 39 (2000) L555–L558.
13. K. Sato, H. Katayama-Yoshida, Ferromagnetism in a transition metal atom doped ZnO, Physica E 10 (2001) 251–255.
14. K. Sato, P.H. Dederics, H. Katayama-Yoshida, Curie temperatures of III–V diluted magnetic semiconductors calculated from first principles, Europhys. Lett. 61 (2003) 403–408.

Mater. Res. Soc. Symp. Proc. Vol. 1183 © 2009 Materials Research Society   1183-FF10-15

Structural and Ferromagnetic Properties of Fe$_3$Si Thin Films Sputter-Deposited on Si(001)

Siao Li Liew, Debbie Hwee Leng Seng, Hui Ru Tan and Dongzhi Chi
Institute of Materials Research and Engineering, A*STAR (Agency for Science, Technology and Research)
3 Research Link, Singapore 117602

ABSTRACT

Fe$_3$Si thin films were deposited on Si(001) from magnetron sputtering with varying deposition rates and subjected to post-deposition anneal (PDA). Structural investigations via XRD, SIMS and cross-sectional TEM reveal that high rate-deposited Fe$_3$Si is amorphous while low rate-deposited Fe$_3$Si is poly-crystalline with significant differences observed at the Fe$_3$Si/Si interfaces. The structural differences were attributed to the influence of deposition rate on the grain nucleation and microstructural morphology in the as-deposited Fe$_3$Si and the subsequent annealed films which in turn determine the ferromagnetic properties. Magnetic properties of Fe$_3$Si deposited at high rate degrade with PDA - coercive field H$_c$ increases from 1 to 14 Oe while saturation magnetization M$_s$ decreases from ~940 to 590 emu/cm^3. In contrast, Fe$_3$Si film sputter-deposited at low rate has a H$_c$ of 5 Oe, M$_s$ of ~920 emu/cm^3 and remnant magnetization M$_r$ ~0.9M$_s$ that are maintained even upon PDA at 350 °C.

INTRODUCTION

The high Curie temperature of 840 K and silicon base of ferromagnetic Fe$_3$Si make it both a practical and compatible candidate for implementation in Si-based electronic devices to be endowed with exciting spin-related functionalities. In particular, epitaxial Fe$_3$Si has been studied often for its potential in spintronic applications due to its high degree of calculated spin polarisation at the Fermi level (DO3 structure) [1,2]. However studies on non-epitaxial Fe$_3$Si and its ferromagnetism-structures relationship are limited presently though the properties of nanocrystalline mechanically alloyed Fe$_{75}$Si$_{25}$ powders have been reported recently [3]. In contrast, other ferromagnetic materials such as polycrystalline Fe$_{1-x}$V$_x$ alloy thin films were already studied for their magnetic properties and magnetization dynamics for application in spin transfer devices [4] and polycrystalline Ni nanowires for the influence of microstructural parameters on the magnetic properties [5]. These studies showed that the magnetic properties of non-epitaxial films could be influenced by controlling the film microstructures.

While it is well understood that the structures of a material are influenced by its processing parameters, the effects of main process variables such as substrate temperature and deposition rates in conventional thin-film processes (e.g. evaporation and sputter deposition) on the structural and ferromagnetic properties of Fe$_3$Si were not investigated before. In this work, we report the structural and ferromagnetic response of Fe$_3$Si thin films sputter-deposited on Si substrates to process conditions, namely deposition rates and post-deposition annealing. The results of this work reveal that judicious choice of deposition rates of Fe and Si are important for thermal stability of structural and magnetic properties of Fe$_3$Si.

EXPERIMENT

Si(001) wafers were cleaned with dilute HF solution prior to film deposition inside a d.c./r.f. magnetron sputtering chamber. To achieve an oxygen-controlled ambience in the chamber during the deposition process, sputtering was started only after the chamber pressure reached at least 3×10^{-7} Torr level. Fe and Si were co-sputtered onto the Si wafers using Ar gas (2.5×10^{-3} Torr) from the respective targets (3" diameter, 0.125" thickness, 4N). One set of samples was prepared with high sputtering rates (0.15 nm/s for Fe and 0.08 nm/s for Si) and another set with low rates (0.027 nm/s for Fe and 0.014 nm/s for Si). Both sets of samples were sputtered to the same total thicknesses of 46 nm and capped with a layer of Si about 5 nm thick to prevent oxidation of the deposited films. The samples with co-deposited Fe-Si layers were subsequently annealed to different temperatures in a *XM80* rapid thermal annealing (RTA) chamber for 60 s in N_2 ambience at 250 and 350 °C to evaluate the thermal evolution of structural and magnetic properties.

Phases and micro-structures were characterized with *Bruker* X-ray area detector diffraction system. Elemental depth compositions were profiled with *ION* time-of-flight secondary ion mass spectrometer (TOF-SIMS). Cross-sectional morphologies of films were examined under *Philips CM300* transmission electron microscope (TEM). Magnetisation curves of films were measured at room temperature with *VSM-880* vibrating sample magnetometer (VSM).

RESULTS AND DISCUSSION

Phase formation

The stoichiometries of as-deposited films for both sputtering rates were first verified with rutherford backscattering spectroscopy (RBS) to be $Fe_{75}Si_{25}$. Figure 1 combines the x-ray diffraction (XRD) plots of Fe-Si films in as-deposited and annealed conditions. Comparing just the as-deposited films first, it is clear that Fe_3Si deposited with high sputtering rate is amorphous (or nano-crystalline with very small grains) while that deposited with low rate is crystalline with

the XRD peak corresponding to $Fe_3Si(220)$ planes at $2\theta \sim 45.3°$ [6]. Rate of deposition $\dot{R}$ influences the final microstructure of film deposited through its effect on the nucleation process variables - critical nucleus size r^* and critical Gibbs free energy change ΔG^* - according to

$\partial r^* / \partial \dot{R} < 0$ and $\partial \Delta G^* / \partial \dot{R} < 0$ (capillarity theory of nucleation [7]). Thus a higher $\dot{R}$ which

implies smaller nucleation islands and lower ΔG^* would favour a nucleation rate $\dot{N}$ that is high

resulting in fine morphology in the film. A lower $\dot{R}$ on the other hand would lead to coarser grained structure due to larger critical nucleus size and nucleation barrier. In the present study, amorphous (fine structure) Fe_3Si were produced under high sputtering rate condition compared to the crystalline form (coarser structure) from low rate condition.

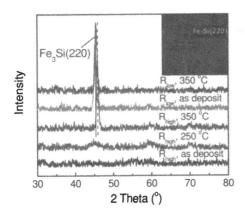

Figure 1. XRD plots of Fe-Si films deposited with different sputtering rates and annealed to various temperatures. (Inset) Area detector scan pattern of low rate-deposited Fe_3Si annealed at 350 °C, showing an arc ring of uniform intensity at 2θ ~45.3°.

Figure 1 also shows that no obvious peak was detected for the high-rate deposited Fe_3Si annealed at 250 °C. The unchanged amorphous structure could be due to the short RTA time though furnace anneal of amorphous Fe_3Si for 300 °C for 10 mins in a separate study did not cause crystallization either. When annealed to 350 °C, the high-rate deposited film became crystalline but probably a different phase from Fe_3Si was formed. This is suggested by the right shift in the 2θ position of the XRD peak to 2θ ~45.9°, relative to the (220) position of Fe_3Si in figure 1. A possible new phase is Fe_2Si [6]. In the case of Fe_3Si films deposited with low sputtering rates, both phase and crystallinity were preserved up to annealing at 350 °C, with the annealed Fe_3Si having similar grain sizes as the as-deposited film. In all the measurements, the angle of incidence was chosen to avoid reflection from Si(004) of the substrate which is close to that of $Fe_3Si(400)$. XRD measurements were repeated at low glazing angles of incidence but were unable to detect additional Fe_3Si peaks besides those from the (220) planes. It could be due to the sensitivity of the XRD machine or the thickness of the films. Nonetheless, the area detector scan pattern showing an arc ring of uniform intensity at the corresponding 2θ angle of $Fe_3Si(220)$ in the inset of figure 1 helps to confirm the poly-crystallinity of Fe_3Si films grown.

Morphological structure

The morphological structures of Fe_3Si were analysed with SIMS and high resolution TEM (HRTEM). SIMS depth profiles in figure 2 show that the Fe_xSi layers were uniform for both as-deposited and annealed samples. Significant interfacial differences were however observed for different rates deposited Fe_3Si. High rate-deposited Fe_3Si has a sharp interface with Si substrate (figure 2a) while annealing to 350 °C led to two pronounced 'humps' in the Fe and Si levels of the SIMS profile (figure 2b). Low rate-deposited Fe_3Si on the other hand, has a less

99

distinct interface as suggested by the pronounced 'hump' in the Si level of the SIMS profile (figure 2c), probably due to a Si-rich layer.

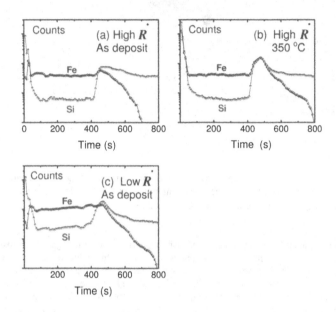

Figure 2. SIMS profiles of Fe-Si films produced under different process conditions. (a) R_{high}, as deposit. (b) R_{high}, 350 °C. (c) R_{low}, as deposit.

HRTEM image in figure 3a of cross-section of as-deposited Fe_3Si (high sputtering rate) shows a relatively sharp interface with Si substrate. The observation of amorphous Fe_3Si is consistent with XRD results shown in figure 1 earlier although localised crystallites in the amorphous matrix could be resolved at higher magnifications. Figure 3b shows an interfacial layer with disordered crystalline atomic arrangement was observed when Fe_3Si film (high sputtering rate) was annealed to 350 °C. The formation coincides with the two pronounced 'humps' in the Fe and Si levels of the SIMS profile shown in figure 2b. RBS measurements indicated that the composition of the thicker overlying Fe-Si layer is Fe_xSi with x ~3 while XRD results mentioned in the preceding subsection on phase formation pointed to Fe_2Si. In contrast, low-rate as-deposited Fe_3Si has an amorphous layer (~1 nm thick) at the interface of Fe_3Si/Si (figure 3c) which most probably corresponds to the Si 'hump' of the SIMS profile.

Figure 3. HRTEM cross-sectional images of Fe-Si films produced under different process conditions. (a) R_{high}, as deposit. (b) R_{high}, 350 °C. (c) R_{low}, as deposit.

The morphological differences between high and low-rate deposited Fe_3Si can be explained by considering that at low rate sputter-deposition, the rate at which the sputtering atoms impinges onto the nuclei ω can have a comparable effect as ΔG^* on $\dot{N}$ (ΔG^* will be higher at low $\dot{R}$ since $\partial \Delta G^* / \partial \dot{R} < 0$).[1] Since ω varies inversely proportional to the square root of the atomic mass [7], $\dot{N}_{Si}$ will be ~1.4 times higher than $\dot{N}_{Fe}$, one therefore expects Si to nucleate

---

[1] From reference [7]:

$\dot{N} = N^* A^* \omega$; N^* - equilibrium concentration of stable nuclei; A^* - critical nuclei area.
$N^* = n_s exp(-\Delta G^*/k_B T)$; n_s - total nucleation site density; ΔG^* - critical Gibbs free energy change; k_B - Boltzmann constant; T - temperature.

first before Fe. This initial preferential nucleation of Si over Fe could be the reason for the observation of a thin amorphous Si layer at the interface of Fe₃Si/Si in the case of low rates co-sputtered Fe and Si. At high rate of sputtering (i.e. reduced ΔG^*), ΔG^* dominates instead due to its exponential effect on N^* and therefore $\dot{N}_{Si} \sim \dot{N}_{Fe}$. The absence of an interfacial layer at the Fe₃Si/Si interface for the high rate-sputtered Fe₃Si film could have allowed inter-diffusion between Fe₃Si layer and Si substrate to occur during annealing which led to the formation of new phase Fe$_x$Si shown in figure 3b.

Ferromagnetism

Figure 4 shows the magnetization behaviours of films prepared under different process conditions with the magnetic fields applied in the plane of the films. Diamagnetic behaviour of bare Si(001) substrate was also measured and the data subtracted from those of the films to exclude the substrate effects.

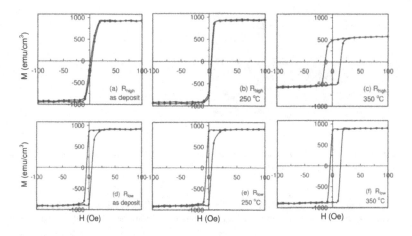

Figure 4. Magnetization curves of Fe-Si films obtained under different process conditions with the magnetic field H applied in-plane to the surfaces of the films. R$_{high}$: (a) as-deposit (b) 250 °C (c) 350 °C. R$_{low}$: (d) as-deposit (e) 250 °C (f) 350 °C.

Figure 4a shows that high rate-sputtered Fe₃Si films exhibit a saturation magnetization M$_s$ of ~940 emu/cm³, coercive field H$_c$ of 1 Oe and remnant magnetization M$_r$ ~0.1M$_s$. XRD and HRTEM results had shown that the Fe₃Si has an amorphous structure. The fine structures of amorphous Fe₃Si would have given rise to the low H$_c$ and squareness ratio M$_r$/M$_s$ that resemble superparamagnetic behaviour [8]. The annealed Fe₃Si at 250 °C exhibited similar ferromagnetic behaviour (figure 4b) as the as-deposited film although it appears that annealing improved the short range order of the magnetic domains in the film as implied by the steeper slope of the magnetization curve at coercivity. When annealed to 350 $^{\circ}$C, the magnetic properties degraded: H_c increased by ten fold to between 14 and 24 Oe while M_s decreased by 35% to ~590 emu/cm^3 (figure 4c). The degradation coincides with the structural changes reported earlier. Grain growth and defects formation such as grain boundaries during crystallization at 350 $^{\circ}$C could have contributed to the increase in H_c. Grain size (D) of a soft ferromagnetic material is known to affect its coercivity (H_c) through $H_c \propto D^6$ [9] in the nanometer grain size regime and increase in H_c had been attributed to grain growth during annealing in soft ferromagnetic thin films [10]. Inter-diffusion between Fe_3Si and Si during anneal and new phase formation Fe_xSi (2<x<3) led to the drop in M_s. The M_s ~590 emu/cm^3 obtained due to this new phase agrees reasonably with reported values for ferromagnetic iron-rich silicides Fe_3Si (970 emu/cm^3) and Fe_5Si_3 (255 emu/cm^3) [11,12].

On the other hand, low rate-sputtered Fe_3Si films display magnetic properties that are thermally stable with M_s ~920 emu/cm^3, H_c 5 Oe and M_r ~0.9M_s (figures 4d to f). The suppressed grain growth after annealing and the presence of a Si-rich interfacial layer help to stabilise the ferromagnetism of Fe_3Si at higher temperature. The H_c values of 5 Oe obtained in this study were amongst the lowest known for Fe_3Si with similar value reported for epitaxial Fe_3Si on Ge(111) [13] and <1 Oe for Fe_3Si/GaAs(001) hybrid layers [14].

It is worth to mention two points at this juncture regarding the ferromagnetic behaviour of Fe_3Si films reported in this study. Firstly, the M_s values for all the as-deposited films compare favourably with those reported by Yoshitake and group for Fe_3Si films on Si(111) [15] but were however lower than that known for bulk Fe_3Si [11] due possibly to the actual stoichiometric ratio of Fe:Si being less than 3. Secondly, the samples generally display displacements of the magnetization curves with varying magnitudes of displacements. Shifts of hysteresis loops are known to occur in cases where a ferromagnetic material in contact with an antiferromagnetic (AFM) is cooled in an external field to a temperature below the Néel temperature of the AFM (field-cooling) [8] with the result that magnetisation is preferred in one direction. In the present study, incorporation of oxygen in the film during sputtering in the deposition chamber could have formed iron oxide that is AFM.

CONCLUSIONS

We have investigated the structural and ferromagnetic properties of Fe_3Si thin films that were sputter-deposited on Si(001) with varying deposition rates and annealed. The magnetic properties of Fe_3Si deposited with high sputtering rate degraded upon annealing with 35% drop in M_s and ten-fold increase in H_c. Fe_3Si deposited with low rate had M_s and low H_c ~5 Oe that were thermally stable. Structural investigations show that high rate-deposited Fe_3Si was amorphous with an almost sharp interface with Si substrate while low rate-deposited Fe_3Si was poly-crystalline with a Si-rich interfacial layer at Fe_3Si/Si interface. The differences were attributed to the influence of sputter-deposition rate on the grain nucleation and microstructure of the as-deposited Fe_3Si and the subsequent annealed films.

103

ACKNOWLEDGMENTS

S. L. Liew wishes to acknowledge the help from B. C. Lim (Data Storage Institute, Singapore) for VSM measurements, H. K. Hui and S. Y. Chow (Institute of Materials Research and Engineering, Singapore) for TEM observations and sample preparation and Prof. T. Osipowicz (National University of Singapore) for RBS measurements.

REFERENCES

1. J. Waliszewski, L. Dobrzyński, A. Malinowski, D. Satula, K. Szymański, W. Prandl, Th. Bruckel , O. Scharpf, *J. Magn. Magn. Mater.* **132** 349 (1994).
2. J. Kudrnovský, N. E. Christensen, O. K. Andersen, *Phys. Rev.* **B43** 5924 (1991).
3. M. P. C. Kalita, A. Perumal, A. Srinivasan, *J. Magnetism and Magnetic Materials* **320** 2780 (2008).
4. J. -M. L. Beaujoura, A. D. Kent, D. W. Abraham, J. Z. Sun, *J. Appl. Phys.* **103** 07B519 (2008).
5. E. P. Hernández, S. M. Rezende, A. Azevedo, *J. Appl. Phys.* **103** 07D506 (2008).
6. Joint Committee on Powder Diffraction Standards (JCPDS): Fe_3Si (45-1207), Fe_2Si (26-1141).
7. M. Ohring, *Materials Science of Thin Films: Deposition and Structure*, 2nd ed., Academic Press, San Diego, Calif., 2002.
8. A. Tomou, D. Gournis, I. Panagiotopoulos, Y. Huang, G. C. Hadjipanayis, and B. J. Kooi, *J. Appl. Phys.* **99** 123915 (2006).
9. G. Herzer, *IEEE Trans. Magn.* **26** 1397 (1990).
10. A. Hashimoto, T. Matsuu, M. Tada, S. Nakagawa, *J. Appl. Phys.* **103** 07E734 (2008).
11. W. A. Hines, A. H. Menotti, J. I. Budnick, T. J. Burch, T. Litrenta, V. Niculescu, K. Raj, *Phys. Rev.* **B13** 4060 (1976).
12. A. N. Hattori, K. Hattori, K. Kodama, N. Hosoito, and H. Daimon, *Appl. Phy. Lett.* **91** 201916 (2007).
13. T. Sadoh, M. Kumano, R. Kizuka, K. Ueda, A. Kenjo, M. Miyao, *Appl. Phys. Lett.* **89** 182511 (2006).
14. J. Herfort, H.-P. Schönherr, K. H. Ploog, *Appl. Phys. Lett.* **83** 3912 (2003).
15. T. Yoshitake, D. Nakagauchi, T. Ogawa, M. Itakura, N. Kuwano, Y. Tomokiyo, T. Kajiwara, and K. Nagayama, *Appl. Phy. Lett.* **86** 262505 (2005).

Multi-ferroic Materials and Oxides

Mater. Res. Soc. Symp. Proc. Vol. 1183 © 2009 Materials Research Society   1183-FF11-01

Magneto-Transport and Magnetic Properties of Fe Doped Nanometric Polycrystalline La$_{0.7}$Sr$_{0.3}$MnO$_3$ CMR Manganites

S. Paul [a) and T. K. Nath
Magnetism and Magnetic Materials Laboratory, Department of Physics & Meteorology, Indian Institute of Technology, Kharagpur 721302, West Bengal, India
a) Corresponding author's email: sanjoy.paul@hotmail.com

ABSTRACT

We have investigated the effect of 3d-transition metal Fe (Iron) doping at Mn site of nanometric polycrystalline La$_{0.7}$Sr$_{0.3}$MnO$_3$ (i.e. La$_{0.7}$Sr$_{0.3}$Mn$_{1-x}$Fe$_x$O$_3$; $0 \leq x \leq 0.1$) CMR manganites on magneto-transport and magnetic properties. Nanocrystalline Fe doped La$_{0.7}$Sr$_{0.3}$MnO$_3$ powders were synthesized through chemical route "Pyrophoric Reaction Process" and calcinated at 850°C for 5 hrs. X-ray diffraction (XRD) patterns of synthesized powder indicate that all samples are having perovskite structure without any secondary impurity phase. Average crystallite size was found to be 20 nm using Debye Scherer formula. Transmission electron micrographs (TEM) show that the average particle sizes are in nanometric regime ($\varphi \sim 50$ nm) and samples are polycrystalline in nature which was observed through selected area electron diffraction (SEAD) patterns. The effect of Fe doping at Mn site of La$_{0.7}$Sr$_{0.3}$MnO$_3$ was found to change substantially the magnetic and transport properties without modifying lattice structure. The suppression of magnetic and transport properties were observed due to dilution of double exchange mechanism in Mn^{3+}- O^{2-}-Mn^{4+} network in La$_{0.7}$Sr$_{0.3}$MnO$_3$.

INTRODUCTION

Magneto-resistive perovskites of the form Re$_{1-x}$Ae$_x$MnO$_3$ (where, Re=La, Nd, Pr, etc. trivalent rare-earth ions and Ae=Sr, Ca, Ba etc. divalent alkaline-earth ions) are subject to extensive research due to their interesting colossal magnetoresistance in recent years [1-2]. Those materials are very useful for the development of magnetic and magneto-resistive devices having special potential application in next generation of digital recording and magnetic sensor [3-5]. Many theories have been proposed to explain the mechanism of CMR effect, such as double exchange (DE) [6], phase separation combined with percolation [7], polaronic effects [8] etc. Numbers of reports have shown that substitution at the A(Re, Ae) or B(Mn) site can significantly modify the magneto-transport and magnetic properties due to the influence of transition metal ions in the double exchange (DE) interaction network (Mn^{3+}-O^{2-}-Mn^{4+}) or through magnetic coupling between the dopant and Mn ions [9-10]. Investigation of magneto-transport and magnetic behaviors in La$_{0.7}$Sr$_{0.3}$Mn$_{0.9}$M$_{0.1}$O$_3$ (M=Cr, Fe, Ni, Cu) nanoscale polycrystals [11], Mn site doping effects of transition elements [12], Effect of Fe doping in La$_{1/2}$Ca$_{1/2}$MnO$_3$ [13] have been extensively investigated.

This paper reports the effect of Fe doping at Mn site of La$_{0.7}$Sr$_{0.3}$MnO$_3$ on magneto-transport and magnetic properties. Two reasons for this work were (a) Fe and Mn have similar atomic radii, and hence the lattice structure will not be modified by Fe doping and (b) Perovskite of La$_{0.7}$Sr$_{0.3}$MnO$_3$ has large band width, so it is unlikely that the phenomenon like electron-phonon interaction characteristics of narrower band system will exist and consequently lattice can be ignored and effect due to change in electronic structure become accessible [13]. So, Fe doping can therefore be used as a control parameter to change only the magneto-transport and magnetic properties.

EXPERIMENTAL

Sample preparation

Nanometric polycrystalline powders of $La_{0.7}Sr_{0.3}Mn_{1-x}Fe_xO_3$ ($0 \le x \le 0.1$) were synthesized through chemical route "Pyrophoric reaction process" using high-purity La_2O_3, Mn $(CH_3COO)_2 \cdot 4H_2O$, Sr $(NO_3)_2$ and Fe $(NO_3)_2 \cdot 9H_2O$, TEA (Triethanolamine, $C_6H_{15}NO_3$) chemicals. We have employed an aqueous solution of the requisite amount of compounds in stoichiometric ratio. TEA was added in this solution, per mole of the total metal ion to form the precursor solution. The amount of TEA was always kept in excess of the total cations present in the mixture. The metal ions to TEA ratio in the starting solutions are maintained at 1:1:4 (i.e. La, Sr: Mn, Fe: TEA:: 1 : 1 : 4) to make a viscous solution. The precursor were then evaporated and pyrolyzed over a hot plate for about 45 min to 1 h at about 180°C with constant stirring. The continuous heating and stirring of the precursor solution caused foaming and puffing. Finally, we got carbonaceous, organic based, black ash. The dried carbonaceous mass was then ground to a fine powder and has been calcinated at a fixed temperature to get a nanocrystalline powders. The heat treatments of the precursor material (in air) were performed for 5 h at 850° C. For magneto-transport and magnetic properties studies nanocrystalline powders were pressed in the form of pellets and re-calcinated at 850°C for 20 min to reduce voids.

Characterization

Structural characterization of the Fe doped nanometric polycrystalline CMR manganites powders of ($La_{0.7}Sr_{0.3}Mn_{1-x}Fe_xO_3$; $0 \le x \le 0.1$) sintered at 850 ° C for 5h, were carried out using X-ray diffraction (XRD) (Models PW 1710 and PW 1810, Philips) with mono-chromatic Cu Kα radiation (λ ~1.542 A°) and by transmission electron microscope (TEM) employing a JEOL 2010F UHR version electron microscope at an accelerating voltage of 200 kV. The peaks of the XRD pattern of the sintered pellets were identified with appropriate hkl reflection following extinction rule for the reported space group (P_{bmn}). Magnetic properties of synthesized powders were done using custom built susceptibility set up to estimate Curie temperature in the temperature range 78-450 K. The DC Resistivity were measured using custom built resistivity measurement set up along with Keithley 2000 digital multimeter (6 and ½ digit) in auto mode and a calibrated Pt-100 temperature sensor attached with a temperature controller (Scientific Instrument Series 5500). Resistance was measured using a standard four probe techniques in temperature range 78-300 K. MR measurement was carried out using the standard Van der Pauw technique at a constant current of 10 mA sample at temperatures 78 K and 300 K, in the transverse geometry (magnetic field is perpendicular to the current, i.e., J $\perp$H) of magnetic field up to 10 kOe. The MR is defined as $(R_H-R_0)/R_0$, where R_H and R_0 are the resistance under applied magnetic field and zero magnetic field.

EXPERIMENTAL RESULTS

Structural characterization

X-ray diffraction patterns show the synthesized Fe doped ($La_{0.7}Sr_{0.3}Mn_{1-x}Fe_xO_3$; $0 \leq x \leq 0.1$) CMR powders (Fig. 1) are having perovskite structure without any secondary or impurity phase. Average crystallites size (D) were estimated using Debye Scherrer formula [$D = k \lambda / \beta_{eff}$ Cos θ, where k is particle shape factor (~ 0.9), λ is the wave length of X-ray (Cu-K$_\alpha$, $\lambda \sim 1.542$ Å), θ is the diffraction angle of the most intense peak (110) centered at around $2\theta = 32.73$, and β_{eff} is defined as $\beta_{eff}^2 = \beta_m^2 - \beta_s^2$, β_m and β_s are the experimental full width at half maximum (FWHM) and the FWHM of a standard silicon sample]. Average crystallites sizes (D) for all synthesized samples were observed to be 20 nm.

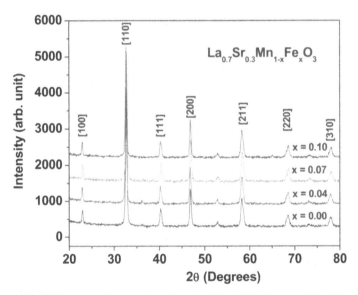

Figure 1 The XRD patterns for samples of $La_{0.7}Sr_{0.3}Mn_{1-x}Fe_xO_3$; $0 \leq x \leq 0.1$) calcinated at 850°C for 5 hrs

Transmission electron micrographs (TEM) (not shown here) show the formation of nanosize CMR particles and average particle sizes are in nanometric regime ($\varphi \sim 50$ nm). Circular rings in selected area diffraction (SEAD) patterns prove the samples are polycrystalline in nature.

Magnetic Properties

Magnetic characterization of all synthesized powders ($(La_{0.7}Sr_{0.3}Mn_{1-x}Fe_xO_3; 0 \leq x \leq 0.1)$) were done using temperature dependent Ac-susceptibility (χ'-T) measurement at frequency 10 KHz. Figure 3 shows the temperature dependent normalized Ac-susceptibility (χ'/χ'_{max}) results for all synthesized powders. It has been observed that the normalized Ac-susceptibility (χ'/χ'_{max}) undergoes sharp transition from ferromagnetic (FM) state to paramagnetic (PM) state. Curie temperature (T_C) is the temperature at which the derivative ($d\chi'/dT$) is minimum. With Fe doping is has been observed that T_C drops from parent compound ($T_C = 358.98$ K) to the CMR material with $x = 0.1$ ($T_C = 264.82$ K) and tunable Curie temperature was observed (Table 1).

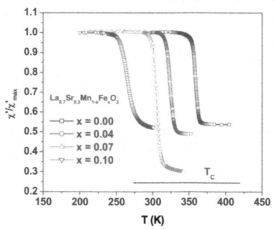

Figure 2 The temperature dependence of the normalized Ac-susceptibility (χ'/χ'_{max}) for $La_{0.7}Sr_{0.3}Mn_{1-x}Fe_xO_3$ ($0 \leq x \leq 0.1$)

Transport Properties

Transport properties of synthesized $La_{0.7}Sr_{0.3}Mn_{1-x}Fe_xO_3$ ($0 \leq x \leq 0.1$) CMR powders have been performed using temperature dependent resistivity (ρ-T) in the temperature range 78-300 K and magnetic field dependent magneto-resistance (magneto-transport) measurement at 78 K and 300 K. The % magneto-resistance (%MR) is defined as [%MR = 100 X$\{R_0- R_H\}$/ R_0) %; where R_H and R_0 are the resistance under applied magnetic field and zero magnetic field]. In temperature dependent resistivity (ρ-T) measurement (Fig. 3a) we observed that the normalized resistivity (ρ_T/ ρ_{max}) increases with T and get maximum at (T=T_P) metal to insulator transition temperature (T_P) and then again decreases with T. It have been observed that the values of T_P decreases with Fe doping from parent ($x = 0$) compound ($T_P = 263$ K) to $T_P = 164.5$ for $x = 0.1$ and tunable T_P observed (Table 1).

In magnetic field dependent magneto-transport studies at T = 78 K (Fig. 3 b) and 300 K (not shown here) we observed interesting negative low field magnetoresistance (LFMR) and very high field MR (HFMR) which have tremendous potential application in low field sensing devices. These curves exhibit the usual behavior of polycrystalline samples with a large (sharp drop) LFMR (H < 2000 Oe) followed by a slower varying MR at a comparatively high field regime (H > 2000 Oe), where MR is almost linear on H. According to Hwang et al. [9] this LFMR in polycrystalline manganites is governed by the spin-polarized transport of conduction electrons across grain boundaries. We defined it as %MR$_{SPT}$. The high field linear %MR (MR$_{INT}$) behavior with H is mainly due to the intrinsic CMR arising from suppression of spin scattering with the increase of field. %MR$_{INT}$ can be calculated by extrapolating the linear part to %MR axis, that is the drop of %MR from low field to high filed. The total contribution in %MR (%MR$_T$) is due to both parts in low and high magnetic field. %MR$_T$ observed for parent sample (x = 0) was observed to be 3.48% at T = 300 K and 21.81% at T = 78 K and decrease of %MR$_T$ were observed with Fe content (x) and %MR$_T$ was observed 1.6% at T = 300 K and 16.68% at T = 78 K for x = 0.1 (Table 1).

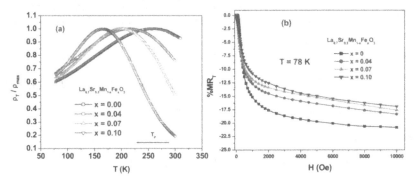

Figure 3 Transport properties of La$_{0.7}$Sr$_{0.3}$Mn$_{1-x}$Fe$_x$O$_3$ (0 ≤ x ≤ 0.1) CMR manganites samples. (a) Temperature variation of normalized resistivity (b) Magneto-resistance measurement at 78 K

Table 1 Table results (T$_C$, T$_P$ and %MR$_T$ at T = 78 K and 300 K) of magneto-transport and magnetic properties of Fe doped nanometric La$_{0.7}$Sr$_{0.3}$MnO$_3$ CMR manganites

x	T_c (K)	T_P (K)	% MR$_T$ at T = 300 K	% MR$_T$ at T = 78 K
0	358.98	263	3.48	20.81
0.04	323.84	222	3.28	18.22
0.07	305.03	201.5	2.68	17.56
0.10	264.82	164.5	1.6	16.68

111

DISCUSSION

Mn site doping has direct impact on the heart of double exchange (DE) interaction where Mn^{3+}-O^{2-}-Mn^{4+} network is the key of this interaction mechanism. Fe doping at Mn site appears to dilute the DE interaction between Mn^{3+} and Mn^{4+} ions in the distorted Mn^{3+}-O^{2-}-Mn^{4+} network and it changes the long-range ferromagnetic ordering as well as the hopping of conduction electrons that possibly explains the observed decrease in T_C, T_P, and $\%MR_T$ with Fe content (x). The observed significant modification in magneto-transport and magnetic properties with Fe doping may also be attributed due to the magnetic spin coupling between Mn and dopant Fe.

CONCLUSIONS

Magneto-transport and magnetic properties of Fe doped properties of chemically synthesized ("Pyrophoric reaction process") $La_{0.7}Sr_{0.3}MnO_3$ CMR manganites (i.e., $La_{0.7}Sr_{0.3}Mn_{1-x}Fe_xO_3$ ($0 \leq x \leq 0.1$) have been investigated through Ac-susceptibility, resistivity and magneto-resistance measurement. The effects of 3d-transition metal (Fe) doping at Mn site in granular nano-crystalline $La_{0.7}Sr_{0.3}MnO_3$ CMR manganites are found to substantially suppress the magnetic and transport properties without changing the lattice structure and explained in the light of dilution of double exchange mechanism.

REFERENCES

[1] R von Helmolt, J. Wecker, B. Holzafel, L. Schultz, and K. Samwer, *Phys. Rev. Lett.* **71**, 2331 (1993)

[2] A. P. Ramirez, *J. Phys.: Condens. Matter* **9**, 8171 (1997)

[3] C. N. Rao and B. Baveau, *Colossal Magnetoresistance, Charge Ordering and Related Properties of Manganese Oxides, World Scientific, Singapore*, 1998

[4] S. W. Li, A. Gupta, G. Xiao, and G. Q. Gong, *Appl. Phys. Lett.* **71**, 1124 (1997)

[5] S. Lee, H. Y. Hwang, B. I. Shraiman, W. D. Ratcliff, and S. W. Cheong, *Phys. Rev. Lett.* **82**, 4508 (1999)

[6] C. Zener, *Phys. Rev.* **82**, 403 (1951)

[7] S. Mori, C. H. Chen, and S. W. Cheong, *Phys. Rev. Lett.* **81**, 5144 (1998)

[8] A. J. Millis, P. B. Littlewood, and B. I. Shraiman, *Phys. Rev. Lett.* **74**, 5144 (1995)

[9] Y. H. Huang, C. H. Yan, L. Zhang, Z M. Wang, C. S. Liao, S. Gao, and G. X. Xu, *J. Appl. Phys.* **90**, 4609 (2001)

[10] J. Balasco, J. Garcia, J. M. de Teresa, M. Ibarra, J. Perez, P. A. Algarabel, and C. Marquina, *Phys. Rev. B* **55**, 8905 (1997)

[11] X. H. Li, Y. H. Huang, C. H. Yan, Z. M. Wang, and C. S. Liao, *J. Phys.:Condens. Matter* **14**, L 177 (2002)

[12] M. M. Xavier Jr., F.A.O. Cabral, J. H. de. Arajujo, C. Chesman, T. Dumelow, *Phys. Rev. B* **63**, 012408 (2000)

[13] P. Levy, L. Granja, E. Indelicato, D. Vega, G. Polla, F. Parisi, *J. Magn. Magn. Mater.* **226-230**, 794-796 (2001)

Appendix

This Appendix contains papers from
Materials Research Society Symposium Proceedings
Volume **1161E**

Engineered Multiferroics — Magnetoelectric
Interactions, Sensors and Devices
G. Srinivasan, M.I. Bichurin, S. Priya, N.X. Sun, Editors

Mater. Res. Soc. Symp. Proc. Vol. 1161 © 2009 Materials Research Society   1161-I01-06

Development of Novel Multiferroic Composites Based on BaTiO$_3$ and Hexagonal Ferrites

D. V. Karpinsky[1], E. K Selezneva[1], I. K Bdikin[1], F. Figueiras[1], K. E. Kamentsev[2], Y. K Fetisov[2], R. C. Pullar[3], J. Krebbs[3], N. M. Alford[3], and A. L. Kholkin[1]
[1]Department of Ceramics and Glass Engineering & CICECO, University of Aveiro, Aveiro, 3810-193, Portugal
[2]Moscow State Institute of Radioengineering, Electronics and Automation, Moscow, Russia
[3]Centre for Physical Electronics and Materials, Department of Materials, Imperial College London, London, UK

ABSTRACT

New multiferroic composite ceramics with the general formula (x)Ba(Sr)Fe$_{12}$O$_{19}$-(1-x)BaTiO$_3$ (x=0.1, 0.5) were synthesized via a simple solid-state reaction technique. Crystal structure analysis performed for such materials revealed the presence of two crystalline phases pertinent to the initial composite components. X-ray diffraction (XRD), scanning electron microscopy (SEM) and atomic force microscopy (AFM) were used to assess the crystallinity, microstructure, and local magnetoelectric interactions between ferroelectric and ferromagnetic grains. Magnetic measurements revealed that the saturation magnetization is proportional to the volume fraction of ferrite phase. Dielectric studies demonstrated strong frequency relaxation due to space charge polarization and high conductivity loss making macroscopic magnetoelectric measurements difficult. Novel nanoscale magnetoelectric effect observed by AFM is discussed.

INTRODUCTION

In the last few years materials which combine ferroelectric and ferromagnetic properties have drawn much attention due to the new physics involved and various potential applications [1-3]. Magnetoelectric (ME) multiferroic materials can change magnetization by applying electric field and *vice a versa* and thus are useful for microelectronic devices due to the potential for controlling the properties by both electric and magnetic fields. A coexistence of ferromagnetism and ferroelectricity rarely coexists in single-phase materials. The problem is mainly caused by the electronic structure of transition ions in ferroelectric compounds which excludes magnetism. As a result, rare ferroelectric materials with magnetic properties basically have an antiferromagnetic structure with low magnetization and magnetoelectric properties are typically observed at low temperatures. An alternate way to combine magnetism and ferroelectricity is to synthesize the composite materials containing separate magnetic and ferroelectric phases. A number of configurations of these components (particulate, layered, core-shell etc.) can produce remarkable ME effects as recently discussed in the literature [4-7].

In this work, we prepared novel magnetoelectric composites based on barium/strontium hexaferrites and barium titanate and investigated their properties. Barium and strontium hexagonal ferrites (abbreviated as BaM and SrM, respectively) have found various applications in the past as a material of choice for permanent magnets and storage media. On the other hand, barium titanate (BT) is a well-known ferroelectric with sufficiently high piezoelectric properties to be used in the composites. Here, we report on the properties of Ba(Sr)Fe$_{12}$O$_{19}$ – BaTiO$_3$ composites with volume fractions 10%-90% and 50%-50% (hereafter abbreviated as BaM-BT 10-90, SrM-BT 10-90, SrM-BT 50-50, and BaM-BT 50-50).

EXPERIMENTAL DETAILS

Composite samples of Ba(Sr)M-BT 10-90 and Ba(Sr)M-BT 50-50) were prepared using a conventional ceramic process. First, stoichiometric amounts of the high purity oxides and carbonates $BaCO_3$, $SrCO_3$, and Fe_2O_3 were mixed in a ball mill, and calcined in air at 1150 °C for about 12 hours, with heating and cooling rates of $5°C$ / min, to produce the ferrites. The ferrite powders were mixed with commercial barium titanate powder (Alfa Aesar) in desired ratios, and ball milled for 24 hours. Then the mixed powder was uniaxially pressed into pellets and finally sintered in air at 1250 °C for 2 hours at heating and cooling rates of 5 °C/min. All heat treatments were carried out in air.

The crystal structure of the samples was analyzed by X-ray diffraction (XRD) technique. XRD patterns were collected at room temperature using a Rigaku D/MAX-B diffractometer with Cu K-α radiation. The data were then analyzed with a Rietveld method [8]. The microstructure of the samples and their local chemical composition was evaluated using SEM (Hitachi S-4100) equipped with an energy dispersive spectroscopy (EDS) detector. The magnetization measurements were performed with MPMS-5 SQUID magnetometer in the temperature range 50-300 K and in fields up to 9 T. The dielectric measurements were done with LCR meters E7-20 and MOTECH 4090D under different test frequencies. Local ferroelectric properties of the samples were analyzed using piezoresponse force microscopy (PFM) method. The measurements were done with a commercial setup (Multimode, NanoScope IIIA, Veeco).

RESULTS AND DISCUSSION

XRD analysis revealed that the crystal structure of the sintered composite samples consists of discrete barium/strontium hexaferrite and barium titanate components. The weight fractions of both phases were close to the initial powders content as verified with Rietveld analysis. Traces of an impurity phase (about 3 wt%) were detected in the XRD pattern of BaM-BT 50-50 sample. The microstructure of the composite samples was analyzed using scanning electron microscopy. The representative SEM picture of the BaM-BT 10-90 sample is shown in the Fig. 1. An EDS analysis revealed that the polyhedral-shaped and elongated grains can be attributed to the ferrite phase whereas rounded ones are ascribed to the ferroelectric fraction. Similar data were obtained for strontium hexaferrite composites. In addition, a wide distribution of grain sizes (400 nm - 4 μm) was observed in all samples. Ferrite grains were substantially bigger than the ferroelectric ones. Such a difference can be explained by the lower sintering temperature for the ferrite component and thus its easier grain growth in contrast to the ferroelectric fraction.

As there is a notable difference in grain sizes for two phases, the total surface area is smaller than it could be in the case of uniform particle distribution. However, this is inevitable for the technique used, and at lower sintering temperatures more equisized phases could be generated, although at the expense of composite density. The good crystallinity and absence of impurity phases in the obtained composite samples could provide a surface area sufficient for magnetoelectric interface interactions between the two fractions. SEM/EDS measurements confirm spatial separation of the phases as well as an absence of significant interdiffusion and formation of undesirable phases detrimental to the properties of composites. In general, SEM/EDS data are in accordance with XRD results and consistent with volume fractions and chemical compositions of starting compounds.

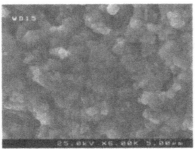

Figure 1. SEM image for the BaM-BT 10-90 sample. Polyhedrally shaped and elongated grains are attributed to the ferrite phase, rounded ones are ascribed to ferroelectric component.

In order to quantify the magnetic subsystem of the composite materials the magnetization measurements were made over a wide temperature range and in the magnetic fields up to 5 T. In general, the magnetic curves revealed a gradual decrease of magnetization with temperature increase without any notable anomaly related to ferroelectric phase transitions in BT. The field dependences of magnetization for the compounds are characteristic for hard magnetic materials (see example in Fig. 2) and are in an agreement with the existing data on parent ferrites [9].

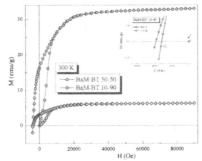

Figure 2. Representative magnetization curves for BaM-BT samples at 300 K. Inset represents the temperature evolution of coercive field for BaM-BT 10-90.

The unusual effect of the increase of coercive field (H_c) with temperature was observed in BaM10-90 samples (inset to Fig. 2). Similar behavior of H_c was found in [10] and can be explained by the change between shape and magnetocrystalline terms in magnetic anisotropy with temperature. Most probably, phase transformation in barium titanate may induce some changes in the grain morphology and thus contribute to the magnetic anisotropy of ferrite component.

Dielectric permittivity and electrical resistivity were studied in a wide range of frequencies (25 Hz - 1 MHz) and temperatures (270 – 520 K). It is known [11, 12] that the ferrite phase has moderate electric resistivity ($\sim 10^8$ Ohm·cm) and thus reduces the total resistivity of the composite making ferroelectric P-E measurements difficult. It should be noted that Sr-containing

117

composites have smaller electrical resistivity than Ba-containing ones. Certain differences in strontium and barium hexaferrite composites are also revealed in the dielectric permittivity data. Though the frequency dependences of permittivity ε demonstrate similar behavior typical for Maxwell-Wagner relaxation the dielectric relaxation for barium ferrite composite is much stronger than for the strontium composite (data not shown here). Such trends are apparently caused by not only intrinsic properties of strontium ferrite but also by notable differences in the microstructures observed in the composites.

Temperature dependences of dielectric permittivity have also been measured for the composites. A diffuse peak is observed for Sr- and Ba-containing composites with 90% barium titanate phase near the Curie temperature of $BaTiO_3$. Different behavior is observed in the composites with 50% of ferroelectric phase, where just a kink in the dielectric curve is observed (Fig. 3) overshadowed by the strong relaxation-type peaks at higher temperatures. This peak can be caused by the Maxwell-Wagner interfacial polarization for these samples. An increase of ε dependence with temperature at low frequencies is revealed for all compounds and it is more pronounced for Ba(Sr)M-BT 50-50 composites. The observed trends are typical for inhomogeneous systems with increasing ac conductivity at higher temperatures.

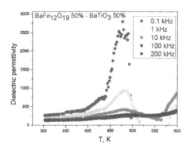

Figure 3. Temperature dependence of dielectric permittivity for BaM-BT 50-50 sample.

As a finite resistivity of the composites prevents macroscopic polarization measurements by regular techniques, their local polarization was studied by atomic force microscopy in piezoresponse force mode (PFM) [7]. Thoroughly polished samples were studied using PFM on ferroelectric grains. Distinct piezoelectric contrast was observed for all composites under study, whereas local piezoelectric hysteresis loops could be successfully measured only on the samples with 90% fraction of ferroelectric phase. The piezoresponse hysteresis loop for SrM-BT 10-90 with subtracted electrostatic interactions is shown in Fig. 4. Specific shape and characteristic parameters indicate a substantial decrease of polarization distribution as compared to pure $BaTiO_3$ ceramics. The results obtained cannot be ascribed to the simple dilution effect by the magnetic phase and other mechanisms should be taken into account. The data obtained testify to a switchable local polarization of the samples under a moderate electric field.

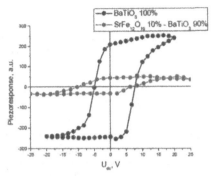

Figure 4. Local ferroelectric hysteresis loops for SrM-BT 10-90 composite and for pure BaTiO₃.

A novel local magnetoelectric effect was observed by measuring magnetic contrast (MFM) of a magnetic BaM grain after local electrical poling of the neighboring ferroelectric BT grain in BaM-BT 50-50. The poling voltage of about 90 V was applied to a conductive tip near a ferroelectric-ferromagnetic grain boundary. The polarization contrast was reversed suggesting that a sufficiently high stress was transmitted through the grain boundary. Then the MFM signal was measured within a ferrite grain. A difference in the cross-sections of MFM before and after poling is shown in Fig.   5.

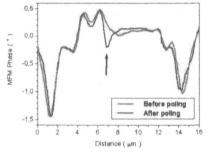

Figure 5. Comparison of MFM signal cross-sections before and after electrical poling for BaM-BT 50-50 sample.

Most probably, the voltage applied resulted in the significant shift of the grain boundary due to electromechanical strain exerted by the tip. This in turn led to the changes in the magnetic interactions in the ferrite grain. The observed strain-mediated response in magnetic signal is evidence of local magnetoelectric coupling between ferroelectric and ferromagnetic components, and can be due to the change of magnetic anisotropy and the apparent width of the domain wall. The calculations verified that the observed effect can be realized if the grain boundary is shifted by only 10 nm. A macroscopic magnetoelectric effect was also measured and yielded a low

119

frequency value of 0.1 mV/cm·Oe for the BaM-BT 50-50 sample. This low value can be a result of the inability to efficiently pole the ceramics, due to the low resistivity of the ferrite component. The work is ongoing to improve the magnetic and ferroelectric phase distribution, so as to achieve higher magnetoelectric coupling while maintaining good ferroelectric and magnetic properties.

CONCLUSIONS

Novel composite ferroelectric/ferromagnetic ceramics were prepared by conventional solid-state sintering. The magnetic properties are consistent with the volume fraction of the magnetic phase present, but ferroelectric properties are degraded via the relatively high conductivity of the ceramics. The macroscopic magnetoelectric effect is rather weak. However, magnetoelectric coupling can be observed at a local level by applying dc voltage to a ferroelectric grain and measuring magnetic response in neighboring magnetic grains. Further improvement of the ferroelectric properties and phase connectivity is required for the use of such composites in practical applications.

ACKNOWLEDGMENTS

The work was partly supported by EC-funded project "Multiceral" (NMP3-CT-2006-032616) and by collaborative project FLAD/NSF 600-06/2006 F. F. and D. K. acknowledge FCT for the support within their grants SFRH/BD/25011/2005 and SFRH/BPD/42506/2007.

REFERENCES

1. J. Wang, J. B. Neaton, H. Zheng, V. Nagarajan, S. B. Ogale, B. Liu, D.Viehland, V. Vatihyanathan, D. G. Schlom, U. V. Waghmare, N. A. Spaldin, K. M. Rabe, M. Wuttig, and R. Ramesh, Science **299**, 1719 (2003)
2. N. A. Hill, J. Phys. Chem. B **104**, 6694 (2000)
3. S. R. Shannigrahi, A. Huang, N. Chandrasekhar, D. Tripathy, and A. O. Adeyeye, Appl. Phys. Lett. **90**, 022901 (2007)
4. Ce-Wen Nan, M. I. Bichurin, Shuxiang Dong, D. Viehland, G. Srinivasan, J. Appl. Phys. **103**, 031101 (2008)
5. G. Liu, C.W. Nan, Z.K. Xu, H.D. Chen, J. Phys D.: Appl. Phys. **38**, 2321 (2005)
6. H.F. Zhang, S.W. Or, H.L.W. Chan, J. Appl. Phys. **104**, 104109 (2008)
7. A. L. Kholkin, S. V. Kalinin, A. Roelofs, A. Gruverman, "Review of ferroelectric domain imaging by Piezoresponse Force Microscopy", in *"Scanning Probe Microscopy: Electrical and Electromechanical Phenomena at the Nanoscale"*, Eds. S. Kalinin, A. Gruverman, Springer, 2006, V. 1, pp. 173-214
8. J. Rodriguez-Carvajal, Physica B. **55**, 192 (1993)
9. S.B. Naranga, C. Singh, Y. Bai, I.S. Hudiara, Mat. Chem. Phys. **111**, 225 (2008)
10. N.A. Frey, R. Heindl, S. Srinath, H. Srikanth, N.J. Dudney, Mat. Res. Bull. **40**, 1286 (2005)
11. I.N. Frantsevich and L.N. Tul'chinskii, Chem. Mater. Sci. **10**, 133 (1971)
12. M.J. Iqbala, M.N. Ashiqa, P.Hernandez-Gomezb, J.M. Munoz, J. Magn. Magn. Mat. **320**, 881 (2008)

Mater. Res. Soc. Symp. Proc. Vol. 1161 © 2009 Materials Research Society 1161-I02-02

Growth of ZnO:Mn/ZnO:V Heterostructures and Ferroelectric-Ferromagnetic
Characterization

Devajyoti Mukherjee, Tara Dhakal, Hariharan Srikanth, Pritish Mukherjee and Sarath
Witanachchi
Department of Physics, University of South Florida, Tampa, Florida

ABSTRACT

The wide band gap semiconductor ZnO is well known for its multifunctionality in the
form of ferromagnetism (FM), piezoelectricity, and magneto optics. ZnO has been found to grow
with intrinsic oxygen deficiencies which in turn are believed to give ferromagnetism and high
conductivity in this material. Doping Zn^{2+} sites by V^{5+} ions creates a mixed valency as well as
strain in the original ZnO hexagonal structure because of the reduced ionic size of vanadium.
The mixed valency creates charge polarity between Zn-O and V-O bonds. This charge polarity
and the rotation of the nonlinear V-O bonds with respect to Zn-O bonds under electric field have
been shown to produce ferroelectricity. Furthermore, Mn doping of ZnO has also shown
enhancement in ferromagnetic properties in ZnO. For this material to be a viable ferromagnetic
material the magnetic properties should not be from segregated phases. In the present study we
have grown undoped, Mn, and V doped ZnO thin films using pulsed laser deposition (PLD).
ZnO target with 2% atomic Mn doping and a target with 0.5% atomic V doping were prepared
by solid state reactions and sintering. Films were grown both epitaxially on sapphire substrates
and in polycrystalline form on silicon substrates. Magnetization measurements by the PPMS
showed M vs. H hysteresis loops with saturation for all ZnO: Mn films. V doped films showed
high saturation polarization for film deposited at high pressures. We have also fabricated
epitaxial bilayers of ZnO:V/ZnO:Mn on sapphire substrates. Ferroelectric and ferromagnetic
properties of these heterostructures are presented.

INTRODUCTION

Room temperature ferromagnetism seen in ZnO is attributed to the intrinsic oxygen
deficiency. Recently, some studies have also revealed the introduction of a strong magnetic
moment to ZnO by doping with Mn [1]. $Zn_{1-x}Mn_xO$ has attracted more attentions because of the
wide band gap of ZnO and the high thermal solubility of Mn in ZnO. Theoretically Mn doped
ZnO is predicted to have the critical temperature (Tc) well above room temperature [2]. Mn
when doped nominally in ZnO is found to be in Mn^{2+} oxidation state and the ferromagnetism is
carrier induced for Mn concentrations below 5 at.% [3]. In contrast to the ref. 3, Kundaliya et al.
[4] have claimed that the ferromagnetism in Mn:ZnO system is due to the metastable phase
(Mn_2O_3) rather than by the carrier induced interaction among substituted Mn ions in ZnO. Our
analysis shows that the ferromagnetism shown at epitaxial Mn doped ZnO induced from carrier
induced interaction and can be explained by using RKKY interaction. Vanadium (V) doped ZnO
has been shown to be ferroelectric [5]. Our goal was to find the growth conditions to maximize
the magnetic moment in Mn doped ZnO and investigate the multiferroic behavior in Mn:ZnO
and V:ZnO heterostructures.

EXPERIMENTAL DETAILS

ZnO:Mn and ZnO:V thin films were deposited on both c-cut sapphire (0001) substrates and Si (100) substrates using pulsed laser deposition (PLD). A pulsed KrF excimer laser (248nm) operating at 10Hz was focused on ZnO:Mn and ZnO:V targets inside a vacuum chamber equipped with a multi target changer sequentially exposing different targets to the laser beam and enabling the in situ growth of hetero-structures with relatively clean interfaces. The laser beam energy was set at 154mJ/pulse. The energy density at the target surface was 2J/cm^2. The target to substrate distance was kept constant at 6cm. The ZnO:Mn and ZnO:V targets were fabricated by conventional solid state reaction by mixing stoichiometric amounts of high purity ZnO,MnO$_2$ powders in case of ZnO:Mn target and ZnO, V$_2$O$_5$ powders in case of ZnO:V target, cold pressing and followed by sintering for 12h at 1000 °C in air. The deposition chamber was pumped to a base pressure of under 10^{-6} T before preparing the thin films and back filling the chamber with high purity oxygen gas. All the films were deposited for 30 mins. The thicknesses of the films were determined using a profilometer. The prepared thin films were measured by X-ray diffraction for crystal structure. All XRD measurements were carried out with Cu Kα radiation. EDS measurements were carried out on the target. The concentration of Mn was found to be 2.2at.% in the ZnO:Mn target and that for V was found to be 0.5 at.% in the ZnO:V target.

RESULTS AND DISCUSSION

Prior to growing the heterostructures, thin films of ZnO:Mn and ZnO:V were grown separately. The growth pressures for ZnO:Mn thin films were varied from 0 mT to 300 mT while growth temperature was chosen to be 600°C. At this growth temperature we observed the best crystallinity in our films. In these ferroelectric/ferromagnetic structures the multiferroic coupling is mediated by the interfacial stress. For this reason films were deposited on c-cut sapphire substrates in order to maintain an epitaxial relation. Films were also grown in polycrystalline form on Si (100) substrates for comparison.

ZnO:Mn thin films

The crystal structure and film orientation of the as-grown films were determined from □-2□ scans of XRD. All XRD patterns indicate that the films are single phase and c-axis preferred oriented. The intensity has been plotted in log scale to emphasize the absence or presence of impurity phases. Figure 1(left) and 1(right) shows the x-ray diffraction scans for the films deposited on Al$_2$O$_3$ substrate and Si (100) substrate at 600°C varying the background oxygen pressure during deposition from 0mT to 300mT respectively. Note that in figure 1 (a) for all the films there was no segregation of the Mn$_3$O$_4$ phase in the diffraction data. The Mn$_3$O$_4$ phase is ferromagnetic with a Curie temperature of less than 50K. However the epitaxial relationship between sapphire substrate and films broke down at higher deposition pressures and other ZnO planes are observed besides the (0002) plane. For the films deposited on Si (100) substrates there was Mn$_3$O$_4$ phase segregation at higher pressures which gave a significant difference in magnetization of these films at 10K.

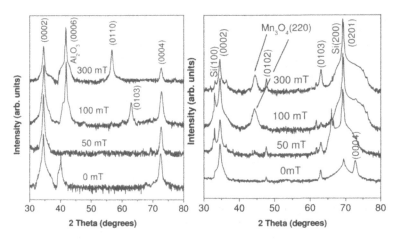

Figure 1. XRD data for Mn doped ZnO thin films deposited at 600°C with varying oxygen background pressure on c-cut sapphire substrates (left) and Si(100) (right) substrates

The magnetization measurements of the thin films were done using Quantum Design Physical Property Measurement System (PPMS) in the ranges of magnetic field (H) from -5T to 5T at temperatures of 10K and 300K. The magnetic field was applied parallel to the film plane. The main hysteresis loops that are reported were obtained after removal of the diamagnetic contribution from the substrates. Figure 2 shows the M vs H loops of ZnO:Mn thin films grown on Al_2O_3 substrates at varying pressures. We observed a drop in the saturation magnetization at 300 K with increasing pressure from 1.47 emu/cm^3 for film deposited at no oxygen background pressure to 0.32 emu/cm^3 for film deposited at 300mT of oxygen pressure. This behavior could be explained on the basis of spin glass behavior as reported by Fukumura et al. [6]. It has been reported that ZnO films deposited at low oxygen pressure have large carrier concentration, with carriers originating from intrinsic defects as Zn interstitials and oxygen vacancies [7]. The less availability of conduction electrons for films deposited at higher O_2 pressures suppresses the magnetic interaction between the Mn^{2+} ions that is mediated by the carriers. This results in a lowering of saturation. Also in order to confirm that Mn doping did play a role in enhancing the ferromagnetic properties of ZnO films hysteresis loops were measured for undoped and Mn doped ZnO films both grown under the same conditions. There has been report of observed room temperature ferromagnetism in undoped ZnO thin films [8]. The saturation magnetization in the Mn doped film was found to be four times higher than undoped ZnO film, both showing room temperature ferromagnetism. (data not presented here)

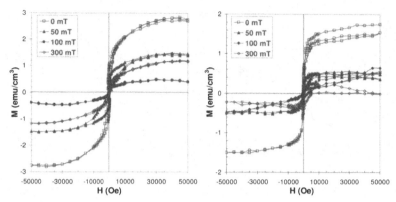

Figure 2. Magnetization loops at 10 K (left) and 300 K (right) for field applied parallel to the film plane for Mn doped ZnO (2 at.% Mn) films deposited at 600°C on sapphire substrates by varying the background oxygen pressure during deposition from 0 mT to 300 mT.

Figure 3 shows the M vs H loops for films deposited on Si (100) substrates at the given conditions. There is a huge change in the sat. mag. at 10 K for the film deposited at 300 mT O_2 pressure. The saturation magnetization for the film deposited at 300 mT is 3.07 emu/cm^3 as compared to 1.83 emu/cm^3 for the 0 mT deposited film. This could be due to cluster formation of Mn_3O_4 phase which is observed in XRD data in figure 1. Apart from the 300 mT film all other films had lower saturation magnetization than those grown epitaxially on sapphire.

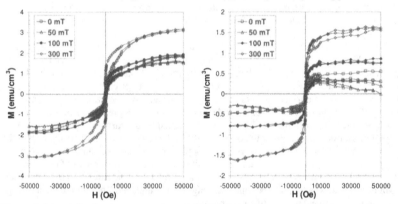

Figure 3. Magnetization loops at 10 K (left) and 300 K (right) for field applied parallel to the film plane for Mn doped ZnO (2 at.% Mn) films deposited at 600°C on Si (100) substrates by varying the background oxygen pressure during deposition from 0 mT to 300 mT.

124

ZnO:V thin films

V-doped ZnO films were grown epitaxially on sapphire substrates at 600°C under different oxygen pressure from 100 mT to 500 mT. XRD data showed good epitaxial relationship with no impurity phases. The films with higher oxygen pressure were more insulating due to less intrinsic oxygen deficiency. Higher O_2 pressure facilitated formation of strong polarity V-O non collinear bonds in V^{5+} substituted Zn^{2+} sites [5]. The saturation polarization was higher for the film grown at higher pressure as shown in figure 4 below. The saturation and remnant polarization are comparable to the values obtained previously [9].

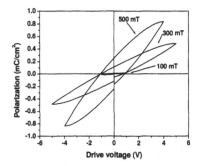

Figure 4. Polarization loops for ZnO:V films deposited on sapphire substrates.

ZnO:Mn / ZnO:V heterostructure

The ZnO:Mn/ZnO:V epitaxial heterostructure was grown as shown in figure 5. The ZnO:V layer was poled by applying 4 V of d.c. bias voltage across coplanar sputter coated Pt electrodes as shown in figure 5.

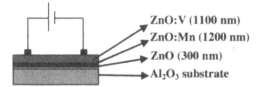

Figure 5. ZnO:Mn/ZnO:V epitaxial heterostructure grown on c-cut sapphire substrate

Figure 6 (left) shows the rocking curve about the (0002) ZnO plane. The FWHM of the rocking curve shows high degree of in plane orientation in the structure. Figure 6 (right) shows the M vs H loops at 300K of the heterostructure before and after poling. Large drop in the sat. mag. by an order of magnitude indicates an interaction between the magnetic moment and the polarization.

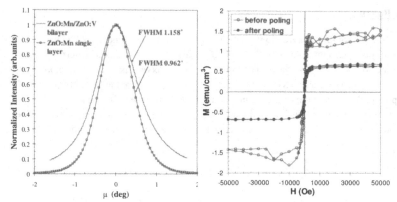

Figure 6. (left) Rocking curves about the (0002) plane of ZnO. (right) Magnetization loops at 300K before and after poling the ZnO:Mn/ZnO:V epitaxial heterostructure.

CONCLUSIONS

In conclusion, room temperature ferromagnetism has been observed in Mn doped ZnO thin films. Magnetization from Mn doped ZnO films grown under various growth conditions suggested that high temperature and low oxygen pressure enhanced the FM behavior. Epitaxial films of V doped ZnO were successfully grown and the ferroelectricity was investigated. The ZnO:Mn/ ZnO:V heterostructure showed possible magnetoelectric coupling at room temperature.

ACKNOWLEDGMENTS

This work was supported in part by a grant from the Department of the Army through USAMRMC grant number W81XWH-07-1-0708.

REFERENCES

1. Z.W. Liu et al., Materials Letters 62, 1255 (2008)
2. T. Dietl, H. Ohno, F. Matsukura, J. Cibert, D. Ferrand, Science **287**, 1019 (2000)
3. Sharma et al, Nature materials, **2**, 673 (2003)
4. Kundaliya et al. Nature Materials **3**, 709 (2004)
5. Y. C. Yang et al., Appl. Phys. Lett. **92**, 012907 (2008)
6. T. Fukumura et al., Applied Physics Letters, 78, 7(2001)
7. S. P. Heluani et al., Thin Solid Films 515 (2006) 2379-2386
8. N. H. Hong, J. Sakai, V. Brize, J. Phys.: Condensed Matter 19(2007) 036219(6pp)
9. Y. C. Yang et al., Appl. Phys. Lett. **90**, 242903, (2007)

Mater. Res. Soc. Symp. Proc. Vol. 1161 © 2009 Materials Research Society   1161-I03-20

Evidence of Magnetoelectric Coupling in Pb(Fe$_{0.5}$Nb$_{0.5}$)O$_3$ Ceramics Through Impedance Spectroscopy, Electromechanical Resonance, and Standard Hysteresis Measurements

Oscar Raymond[1], Reynaldo Font[1,2], Jorge Portelles[1,2], Nelson Suárez-Almodovar[1,2], and Jesús M. Siqueiros[1]
[1]Centro de Nanociencias y Nanotecnología, UNAM, Ensenada, Baja California, México, 22860
[2]Facultad de Física, Universidad de la Habana, San Lázaro y L, 10400, La Habana, Cuba

ABSTRACT

Multifunctional materials such as the single phase compound Pb(Fe$_{0.5}$Nb$_{0.5}$)O$_3$ (PFN), where ferroelectric and antiferromagnetic order coexist, are very promising and have great interest from the academic and technological points of view. In this work, coupling of the ferroelectric and magnetic ordering has been observed. For this study, a combination of the small signal response using the impedance spectroscopy technique and the electromechanical resonance method with the large signal response through standard ferroelectric measurement, has been used with and without an applied magnetic field. The measurements to determine the electrical properties of the ceramic were performed as functions of the bias and poling electric fields. A simultaneous analysis of the complex dielectric constant $\tilde{\varepsilon}$, impedance $\tilde{Z}$, electric modulus $\tilde{M}$, and the electromechanical coupling factors is presented. The results are correlated with a previous study of structural, morphological, small signal dielectric frequency-temperature response, and the ferroelectric hysteretic, magnetic and magnetodielectric behaviors. The observed shifts of the resonance and antiresonance frequency values can be associated with change of the domain size favored by the readjustment of the oxygen octahedron when the magnetic field is applied. From P-E hysteresis loops obtained without and with an external applied magnetic field a maximum value of dc magnetoelectric coefficient α_{ME} = 4 kV/cm T (400 mV/cm Oe) was obtained.

INTRODUCTION

In many ferroelectric compounds with perovskite structure [A$'$A$''$(B$'$B$''$)O$_3$] in which electric and magnetic order coexist, the study of magnetoelectric coupling, understood as the coupling between both orderings where a change in the ferroelectric state or an external electric field induce a change in the magnetic properties, or vice versa, a change in the magnetic order or an external magnetic field induce a change in the electric properties, is very interesting from both the points of view of materials science and technological applications as new devices such as non volatile memory and/or ferroelectromagnetic devices [1-5].

Single phase multiferroic lead iron niobate Pb(Fe$_{0.5}$Nb$_{0.5}$)O$_3$ (PFN in reduced notation), is a ferroelectric and antiferromagnetic compound with high dielectric constant where Pb^{2+} in site A and Nb^{5+} in site B$''$ favor electrical order and the magnetic moment of Fe^{3+} in the B$'$ site is the source of magnetic order. A detailed study of the structural and morphological properties, magnetic measurements, and of the dielectric properties as functions of temperature and frequency as well as the dielectric relaxation processes near the transition temperature using impedance spectroscopy (IS) of ferroelectromagnetic PFN ceramics, obtained at 900 °C from different precursors using the conventional ceramic method, was reported in previous works [6-

8]. Attending to their better dielectric and ferroelectric properties, the samples here investigated were obtained using the $FeNbO_4$ precursor with orthorhombic structure [6].

The samples showed good crystalinity, exhibiting rhombohedral structure at room temperature (RT) [9], uniformly distributed grains, and a ferroelectric–paraelectric non-relaxor diffuse phase transition at 383 K (110 °C) characterized by a tetragonal-to-cubic symmetry change. Measurements of magnetic susceptibility exhibited a antiferromagnetic-paramagnetic transition above the Néel temperature reported near 122 K, and confirmed by electron paramagnetic resonance (EPR) spectra at 145 K [6,10]. Moreover, from a combination of standard ferroelectric and EPR measurements at cryogenic temperatures, magnetodielectric coupling effects were observed in the paramagnetic to weakly magnetized antiferromagnetic transition near 103 K associated with a rhombohedral to monoclinic symmetry reduction [10]. However, ac magnetoelectric (ME) effects measured through the magnetoelectricity coefficient have not been observed yet.

Here, to investigate the magnetoelectric coupling in this single phase compound, we report a study of the ferroelectric and magnetic ordering coupling from a combination of the small signal response using the IS and the electromechanical resonance method with the large signal response through standard ferroelectric measurement, with and without an external magnetic field. Magnetoelectric coupling effects have been observed and are discussed in correlation with the previous studies.

EXPERIMENTAL DETAILS

A detailed description of the fabrication process of the samples here investigated and the experimental procedures used for the crystallographic, compositional and surface morphological studies, the temperature-frequency response characterization, and the magnetic study is reported in previous works [6-8, 10]. The samples have disc shape with 11 mm in diameter and 1 mm in thickness. The electrical measurements were taken in the 20 Hz to 1 MHz frequency range, with an applied voltage of 1V, using an HP4284 LCR meter. Measurement of the electromechanical resonance were realized, in a 4 HZ – 4 MHz frequency range, using a resonance circuit and a spectrum analyzer built according to IEEE standards for piezoelectric materials. Polarization vs. electric field (P-E) loops were measured using a Precision Ferroelectric Tester system by Radiant Technologies Inc. (in a virtual ground mode). All measurements were realized at room temperature using a two electrode sample holder inserted, for measurements with the magnetic field, into a 1 cm gap of a system built with 4 axially magnetized NdFeB magnetic discs. The dc magnetic field H = 0.76 T was tested with a GM05 Gaussmeter with resolution of 1 mT.

DISCUSSION

As mentioned above, an ac ME effect has not been observed in the PFN compound from measurements of the ME voltage coefficient $\alpha_{ME} = \delta E/\delta H$ which can be measured for longitudinal (all the fields parallel to each other) or transverse (H and δH perpendicular to E and δE) fields. Here, we investigate the longitudinal ME coupling effects from the electrical or electromechanical response of this single phase multiferroic material both under a small or large signal electric field when the external magnetic field is applied parallel to the electric fields.

Small signal ME coupling effects.

Figure 1 show the frequency dependence of some complex functions with and without the external applied magnetic field H. As can be seen in Figure 1a, whereas the value of the real part ε' of the relative dielectric constant decreases with H, the imaginary part ε'' increases. Moreover, whereas the difference of the values of ε' is constant in all frequency range, the difference of ε'' values decrease when the frequency increases and turn zero at frequencies higher than 10^4 Hz, i. e., the effect of $\Delta\varepsilon''$ is more pronounced at low frequencies. From previous studies using the IS [6-8], this behavior can be due to the interaction of the magnetic field with the localized polarization mechanism associated to n and/or p hopping charge related to Fe ions presence in the oxygen octahedron. These interactions promote the disordered movement of magnetic moments increasing the dielectric loss and subsequently the decrease of the polarization. However, from the independence of ε'' with H at high frequency, it can be observed that such interaction is not associated to the long-range conductivity mechanism. This statement agrees with the behavior illustrated in Figure 1b where the real Z' and imaginary Z'' parts of the impedance do not show any dependence with H in all frequency range, and in Figure 1c where the electric modulus does not show substantial difference at higher frequencies.

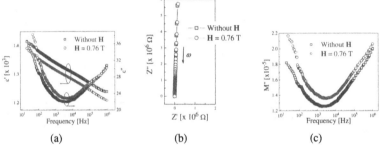

(a)   (b)   (c)

Figure 1. Frequency dependence of some complex functions with and without an external applied magnetic field H. (a) Plots of the real ε' and ε'' imaginary parts of the relative dielectric constant, (b) Nyquist diagrams of the impedance (Z'' vs Z'), and (c) plots of the imaginary part M'' of the electric modulus.

The electromechanical resonance response of the poled PFN ceramic samples in the frequency range corresponding to the fundamental radial mode is illustrated in Figure 2. The samples were electrically polarized in silicon oil (following the standard poling method) at 100 °C (tetragonal ferroelectric phase) under a poling field perpendicular to the flat faces of the discs of 6.3 kV/cm above the coercive field E_C (see below), and at 120 °C (cubic paraelectric phase) with a poling field of 2.5 kV/cm. A shift at higher values of the resonance f_r and antiresonance f_a frequencies is observed for both samples when the magnetic field is applied. Values of the differences Δf_r and Δf_a are reported in Figures 2a and 2b. This shift effect is more pronounced for the sample poled at 100 °C and can be attributed to a better domain structure reached during the poling process. This is reflected in the higher value of the electromechanical coupling factor k_p (here independent of the magnetic field) of 0.21 for the sample poled at 100 °C with respect to the value of 0.15 for the sample poled a 120 °C.

In order to make a better evaluation of the interaction of the magnetic field with the ferroelectric order, measurements of resonance response were realized on a non poled sample but where a bias electric field (E_{BIAS}) perpendicular to the flat faces is applied. Figure 3a illustrates the electromechanical resonance spectra obtained at different values of E_{BIAS}; in the inset, the spectrum of a poled PZT standard sample is presented as reference of validation of our resonance system. The behavior of the fundamental radial mode is exhibited in Figure 3b at different values of E_{BIAS} between 5 and 30 kV/cm without H. A non linear increase of the maximum amplitude and a shift of f_r and f_a frequencies to higher values with the increase of E_{BIAS} is observed in Figure 3b as a consequence of a better orientation of those domains which are possible to switch in the direction of field. As other evidence of the ME coupling, Figure 3c shows the increase in the signal amplitude and the shift of f_r and f_a frequencies when the magnetic field is applied. This effect can be associated to the change of the domain size favored by the readjustment of the oxygen octahedrons when the applied magnetic field acts on the magnetic moment of the Fe ions in these sites.

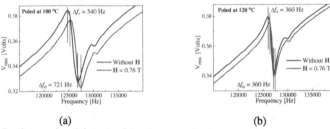

(a)   (b)

Figure 2. Fundamental radial mode of the electromechanical resonance spectra obtained with and without an external applied magnetic field H for samples poled at (a) 100 °C with 6.3 kV/cm, and (b) at 120 °C with 2.5 kV/cm.

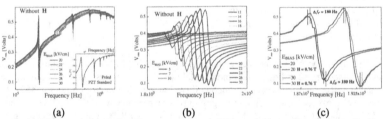

(a)   (b)   (c)

Figure 3. (a) Electromechanical resonance spectra at different values of bias electric field (E_{BIAS}) corresponding to a non-poled sample; inside the spectrum of a poled PZT standard sample as reference. (b) Fundamental radial mode behavior at different values of E_{BIAS}. (c) Comparison of the fundamental modes obtained under E_{BIAS} = 20, 30 kV/cm with and without an external magnetic field H.

Large signal ME coupling effects.

130

Finally, Figures 4a and 4b exhibit the P-E hysteresis loops obtained for different values of the maximum electric field E_{max} without and with an external applied magnetic field, respectively. The dependence on E_{max} of the characteristic parameters of the loops i. e. maximum polarization P_{max}, coercive field E_C, and remanent polarization P_r, is illustrated in Figure 5. As can be observed, the values of P_r (Figure 5a) are not affected by the magnetic field, however, noticeable differences are shown in Figures 5b and 5c for the values of E_C and P_r which are higher when the magnetic field is applied. From the values of P_r in Figure 5c, dc ME coefficients $\alpha_{ME} = \Delta E / \Delta H$ were calculated with a maximum value of $\alpha_{ME} = 4$ kV/cm T (or 400 mV/cm Oe) at $E_{max} = 23$ kV/cm. The P-E loops were obtained at 60 Hz, then in correspondence with the arguments discussed above, the interaction of the magnetic field with the localized polarization mechanism in combination with the high values of the electric field lead to a new redistribution of domains aligning in the direction of the electric field increasing, therefore, the remanent polarization by a higher pinning of domains and, consequently, an increase of the coercive field necessary to remove the pinning and switch the domains. This effect can be strongly increased for higher magnetic fields [11].

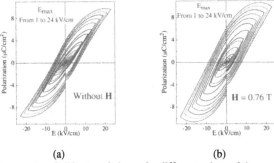

(a)   (b)

Figure 4. Room temperature P-E hysteresis loops for different values of the maximum electric field E_{max}, (a) with and (b) without an external dc magnetic field H.

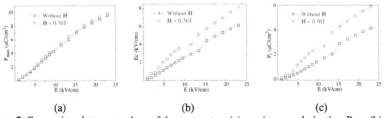

(a)   (b)   (c)

Figure 5. Comparison between values of the parameters (a) maximum polarization P_{max}, (b) coercive field E_C, and (c) remanent polarization P_r, as functions of maximum electric field E_{max}, from the P-E loops obtained with and without an external applied magnetic field H.

131

CONCLUSIONS

In summary, ME coupling effects have been observed in PFN ceramics from measurements of electrical properties under small and large electric signal when an external magnetic field is applied. A change in the ferroelectric order has been induced by means of an external magnetic field and can be correlated with the interaction of this magnetic field with the localized polarization mechanisms and with the magnetic moment of the Fe ions in the oxygen octahedrons.

ACKNOWLEDGMENTS

This work was partially supported by DGAPA-UNAM (Projects IN109608 and IN102908) and CoNaCyT (Projects Nos. 49986-F and 82503). The authors thank E. Aparicio, P. Casillas, J. Peralta, J. Hernández, A. Tiznado, and E. Medina for their technical assistance.

REFERENCES

1. R. Ramesh and N. A. Spaldin, Nature Materials 6, 22 (2007).
2. J. F. Scott, Nature Materials 6, 256 (2007).
3. M. Gajek, M. Bibes, S. Fusil, K Bouzehouane K, J. Fontcuberta, A. Barthelemy, A. Fert, Nature Materials 6, 296 (2007).
4. W. Eerenstein, M. Wiora, J. L. Prieto, J. F. Scott, and N. D. Mathur, Nature Materials 6, 348 (2007).
5. Y. H. Chu, L. W. Martin, M. B. Holcomb, R. Ramesh, Materials Today 10, 16 (2007).
6. O. Raymond, R. Font, N. Suárez-Almodovar, J. Portelles, J. M. Siqueiros, J. Appl. Phys. 97, 084107 (2005).
7. O. Raymond, R. Font, N. Suárez-Almodovar, J. Portelles, J. M. Siqueiros, J. Appl. Phys. 97, 084108 (2005).
8. O. Raymond, R. Font, N. Suárez-Almodovar, J. Portelles, J. M. Siqueiros, J. Appl. Phys. 99, 124101 (2006).
9. M. E. Montero-Cabrera L. Fuentes-Montero, L. Calzada, M. Pérez De la Rosa, O. Raymond, R. Font, M. García, A. Mehta, L. Fuentes., Integrated Ferroelectric 101, 101 (2008).
10. R. Font, G. Alvarez, O. Raymond, J. Portelles, J. M. Siqueiros, Appl. Phys. Letters 93, 172902 (2008).
11. W. Prellier , M. P. Singh, P. Murugavel, J. Phys.: Condens. Matter 17, R803–R832 (2005).

Mater. Res. Soc. Symp. Proc. Vol. 1161 © 2009 Materials Research Society   1161-I03-22

Ultraviolet Photoconductivity of Pure and Al Doped ZnO Thin Films by Inkjet Printing

Yan Wu, Takahiko Tamaki, Wolfgang Voit, Lyubov Belova, K. V. Rao

Department of Materials Science-Tmfy-MSE, Royal Institute of Technology

S-100 44 Stockholm, Sweden

ABSTRACT

Pure ZnO, and Al doped ZnO, 120 -300 nm thin films on glass substrates, were synthesized by inkjet printing technique using zinc and aluminum acetate solution as precursors and a two stage heat treatment process to obtain polycrystalline hexagonal wurtzite structure with the mean grain size of 25 and 30 nm respectively. All films exhibit a transmittance above 85-90% in the visible wavelength range below 700 nm. In the Al doped films the UV absorption spectra show a strong absorption onset below 380nm followed by shoulders centered around 325 nm depending on the film thickness. The electrical conductivity of Al doped ZnO thin films is larger by two orders of magnitude than that for pure ZnO films while the photoconductivity increases by about three orders of magnitude under UV irradiation. The photoresponse of the films with UV irradiation in terms of the rise and decay times in the frequency range from 5 to 500 Hz is also presented and discussed.

INTRODUCTION

Ink jet printing techniques provide a simple low-cost method to use a solution-process at ambient temperature conditions for developing thin film semiconductors and devices [1-3]. This technique is being extensively used today for various applications in the currently active areas of research such as integrated circuits, transparent electronics, and development of ceramic components, biotechnology, organic light-emitting diodes, and polymers, in addition to conventional graphics applications.

Low-cost, nontoxic, environmentally friendly, highly sensitive ultraviolet (UV) sensors are of great interest for the detection of UV light in various special environments. In particular, as a wide band gap (3.3 eV at 300 K) semiconductor with large exciton binding energy (60 meV), ZnO has been intensively studied for such device development [4-6]. The electrical conductivity of pure ZnO thin films is due to either oxygen deficiencies or presence of interstitial Zn in the ZnO lattice. Any additional carrier concentration from a trivalent dopant, (eg. In^{3+}, Al^{3+}, Ga^{3+} replacing Zn^{2+}) increases the electrical conductivity of the ZnO thin films. However, it is well recognized that the measured electrical conductivity highly depends upon the film deposition technique, crystalline quality of the annealed films, heat treatment schedule, and annealing ambience. Thus, the fabrication of a highly sensitive, inexpensive and large-scale UV sensor using ZnO-based films continues to be a challenge. Although there exist some studies on photoconductivity in pure ZnO, reports on UV photoconductivity in Al doped ZnO films [7-9] are rare, and insufficient to exploit satisfactorily the UV response characteristics. In this paper we present what is perhaps the first early study of UV photoconductivity of pure ZnO and Al doped ZnO (AlZnO) polycrystalline 120 to 300 nm thin films fabricated by inkjet printing technique. The films have been carefully characterized using an ultra-high resolution SEM/FIB Dual Beam instrument, and we report our preliminary investigations for their UV response at 50 Hz using a suitable custom built circuitry.

EXPERIMENTAL DETAILS

ZnO and Al doped ZnO precursor inks were prepared for inkjet printing as follows: (1) A 0.25 M (moles/l) ZnO precursor solution was prepared by dissolving zinc acetate dihydrate (ZnAc, Fisher Chemicals) in isopropoxy ethanol (Sigma-Aldrich). The molar ratio of 2-ethanolamine (Sigma-Aldrich) to ZnAc was maintained at 1.0. The final ZnO precursor ink was thus obtained after ultrasonic stirring for 20 minutes at room temperature. (2) For Al doped ZnO precursor ink, a 0.25 M (moles/l) of aluminum nitrate hydrate ($Al(NO_3)_3$) and zinc acetate dihydrate (ZnAc, Fisher Chemicals)(the molar ratio of Al to (Al +Zn) is 2%) was taken in the solution in isopropoxy ethanol (Sigma-Aldrich). The molar ratio of 2-ethanolamine (Sigma-Aldrich) to ZnAc and $Al(NO_3)_3$ is 1.0. This was then followed, as before, by ultrasonically stirred for 20 minutes at room temperature.

The precursor solution was then inkjet printed on glass substrates pre-heated at 50 °C, subsequently dried on a hot plate at 80 °C to evaporate the liquid solvent. This is followed by additional layers deposited and heat treated in a similar manner after each deposition. We designate one layer as one pass in the processing. In our studies we use films with multiple passes. Then the as-deposited final films were processed in the preheated oven at 400 °C for 3 hours to form dense polycrystalline films for further characterization and measurements.

The crystalline structure of films was determined by X-ray diffractometer (Siemens D5000) with Cu $K\alpha$ radiation source in the range of 20 to 70° with 0.02° step. Film thickness was determined by a Dual Beam FIB/SEM (Nova 600 Nanolab by FEI) instrument from the cross-section analysis to a precision of ± 5 nm. The optical transmission spectrum was measured over the 200-700 nm wavelength range using Perkin Elmer Launch 900 optical spectrophotometer equipped with an integrating sphere detector. The electrical properties were measured by four-probe point technique controlled by computer software. The photoresponse was measured by custom-designed setup (using copper contacts in a coplanar gap (0.5 mm) configuration shown in the schematic figure 5) using high precision Keithley measurement instruments, by irradiating with light emitted diode (wavelength 363 nm and optical power 1.5 mW at forward current 20 mA). The UV response measurements were performed using synthesized functional analyzer in the frequency range from 5 to 500 Hz and with a high-resolution oscilloscope (Tektronix 7503).

RESULTS AND DISCUSSION

Crystalline structure characterization and microstructure of the films

Figure 1 shows the X-ray diffraction patterns of pure and Al doped ZnO films. The intensity peaks identified using the JCPDF file for ZnO correspond to the (100), (002), (101), (102), (110), (103), and (112) reflections for wurtzite type ZnO, and we do not find any other peaks which could be identified with impurities or other phases. These results suggest that 400 °C is sufficient for thermal decomposition of $Al(NO_3)_3$, $Zn(Ac)_2$ to crystalline zinc oxides and Al substitutes for Zn in the ZnO structure. The grain size of these samples can be estimated using the Scherrer's formula [10],

$$D = \frac{0.9\lambda}{B\cos\theta} \quad (1)$$

Where, λ is the X-ray wavelength (Cu Kα radiation), θ is the Bragg diffraction angle, and B is the full width at half maximum. Three strongest diffraction peaks in the pattern, (100), (002) and (101) are chosen to calculate the mean grain size. The mean grain size of the films is obtained as 25 and 30 nm respectively in the ZnO and AlZnO films.

The surface morphologies of AlZnO thin film of 5 passes deposition are shown in figure 2. The film exhibits a porous microstructure with a crystalline particle size of approximately 30-50 nm. The film thickness determined from the FIB cross section analyses is about 145 nm (see figure 2 (B)). The film thickness uniformity is found to be within a 5 nm range.

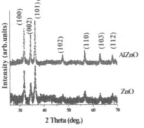

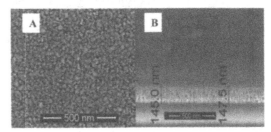

Figure 1. XRD patterns of ZnO and AlZnO films

Figure 2. SEM/FIB images of AlZnO film: (A) top surface image; (B) FIB Cross section image.

Optical properties

We have investigated quantitatively the transparency of thin films through optical transmittance measurements in the near UV and VIS region. Based on the characteristic of inkjet printing, the thinnest continuous film in this study is obtained for a thickness of around 120 nm. We therefore investigated films in the thickness range of 120 to 300 nm.

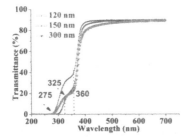

Figure 3. Optical transmittance spectra of AlZnO thin films with 120 nm, 150 nm, and 300 nm thickness.

Figure 3 shows the transmittance spectra in the wavelength range 200-700 nm of AlZnO films for various thicknesses. The average transmittance is found to be more than 85% for VIS light and on an average much less than 20% in the UV range 275 to 325 nm (In the spectra shown, affect of the absorption due to the substrate has not been accounted for). Clearly, the absorption characteristics are found to be thickness dependent. The observed differences in the UV absorption could be possibly due to the porosity and surface roughness formed during the multiple printing-passes, and the affect of the substrate. The observed steep absorption edge at 360 nm with a little broad pre-edge between 275 and 360 nm, is similar with the results observed on films produced by sol-gel processes [11]. Thus our preliminary studies on ink jet films show the well known characteristics of UV light absorption and can be tailored in doped and undoped films. There is indeed much room for improving and perfecting our technique to obtain better absorption characteristics from these films.

Electrical conductivity of ZnO and AlZnO films

The conductivity measurements were made with a four terminal technique and all the current-voltage characteristics were linear. The conductivity dependence vs. the film thickness of pure ZnO and AlZnO are shown in figures 4(A) and 4(B) respectively. As shown in Figure 4 (A), increasing the ZnO film thickness from 120 to 150 nm also increases the electrical conductivity measured in ambient light from $7.44*10^{-6}$ to $2.87*10^{-5}$ ($\Omega \cdot cm)^{-1}$. Under UV illumination the conductivities also increases with increasing ZnO film thickness but from $5.7*10^{-3}$ to $1.6*10^{-2}$ ($\Omega \cdot cm)^{-1}$ or about three orders of magnitude higher as compared with that in the ambient light. In both irradiation processes the conductivity values begin to fall of gradually with increasing film thicknesses above 150 nm.

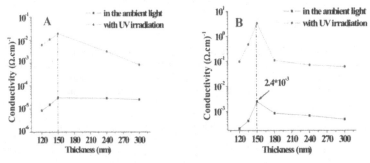

Figure 4. UV lights irradiation effect of the conductivity as a function of film thickness for (A) pure ZnO and (B) AlZnO films

Aluminum doping enhances, in general, the electrical conductivity of ZnO films. Figure 4 (B) shows the electrical conductivities of AlZnO films in the room light and under UV irradiation. The conductivity again initially increases almost linearly from $2.10*10^{-4}$ to $2.40*10^{-3}$ ($\Omega \cdot cm)^{-1}$ when the film thickness increases from 120 nm to 150 nm, which could be explained in terms of the films porosity and the possibility that the films become denser when two more passes are added. The conductivity observed under UV irradiation increases from 0.1 to $3.3(\Omega \cdot cm)^{-1}$ when the film thickness increases from 120 to 150 nm. These results are comparable to those reported by A.E.J.Gonzalez[7], in which the dark conductivity of the AlZnO films with the thickness 167.8 nm is obtained as $3.17*10^{-3}$ ($\Omega \cdot cm)^{-1}$ while the photoconductivity is enhanced to 6.35 ($\Omega \cdot cm)^{-1}$. Notice that the conductivities under UV illumination have increased by about three orders of magnitude compared with those observed in ambient light in our experiment. The increase in electrical conductivity brought by the aluminum doping can be explained as follows: The concentration of free charge carriers in ZnO increases by the aluminum doping because aluminum has one valence electron more than zinc. We may consider that aluminum substitutes the zinc atom or it occupies interstitial positions [12].

UV pulse response of pure ZnO and Al doped ZnO films

Studies of the photoresponse of ZnO UV detectors have shown that the response consists of two main parts: a fast process due to the electron–hole pair generation and a slow process mainly attributed to the oxygen chemisorptions as well as the grain boundaries [2, 13-15].

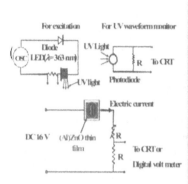

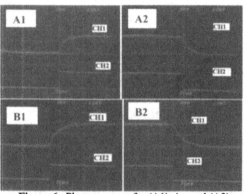

Figure 5. Schematic of electric circuit for photoresponse easurement

Figure 6. Photoresponse for (A1) rise and (A2) decay part of ZnO, (B1) rise and (B2) decay part of AlZnO. Photoresponse CH1 (Horiz. 10 ms/div. Vert. 5mV/div), UV excitations CH2 (Horiz. 10 ms/div. Vert. 20mV/div)

In the current study, only the rise and decay characteristics of the UV response at 50 Hz for the doped and undoped ZnO are discussed. We will report elsewhere more detailed studies in the frequency range from 5 Hz to 500 Hz. Figure 5 is the schematic of electric circuit for photoresponse measurement. The device response in terms of rise and decay parts for ZnO and AlZnO films with the same thickness of 150 nm was studied, while the square pulse UV irradiation waveform is 20 milliseconds for one period of 50 Hz. Figure 6 (A1, A2, B1 and B2) shows the rise and decay times for ZnO films, the rise and decay times for AlZnO film with the pulse UV irradiation, respectively. The rise time observed is in a range of few milliseconds while the decay time is slightly longer for pure ZnO film. Thus, after doping with Al, the rise time increases to about ten milliseconds and the same is true for the decay time. Both of them are nearly twice longer than those of pure ZnO film.

CONCLUSIONS

We have inkjet printed polycrystalline transparent pure ZnO and AlZnO thin films with average grain sizes 25 and 30 nm, respectively. All films exhibit transmittance above 85-90% in the visible wavelength range below 700 nm. Al doped films the UV absorption spectra shows a strong absorption onset below 380nm followed by shoulders centered around 325 nm depending on the film thickness. Detailed analyses of the optical absorption will be reported elsewhere.

The electrical conductivites of ZnO films in the ambient light initially increase from $7.44*10^{-6}$ to $2.87*10^{-5}$ ($\Omega \cdot cm$)$^{-1}$ when the film thickness increases from 120 to 150 nm, while the values of the films under UV irradiation increase from $5.7*10^{-3}$ to $1.6*10^{-2}$ ($\Omega \cdot cm$)$^{-1}$. That shows the conductivities under irradiation have increased by about three orders of magnitude, compared with the ambient light. But the values begin to fall when the thickness increase further. Aluminum doping enhances two orders of magnitude the electrical conductivity of ZnO films. When the film thickness increases from 120 to 150 nm, the ambient light conductivity initially increases from $2.10*10^{-4}$ to $2.40*10^{-3}$ ($\Omega \cdot cm$)$^{-1}$, while the values of the films under UV light irradiation increases from 0.1 to 3.3 ($\Omega \cdot cm$)$^{-1}$. Obviously, the conductivities under irradiation have increased by about three orders of magnitude compared with in the ambient light, and then the values begin to fall gradually for the thicker films.

The photoresponse for ZnO and AlZnO films with the thickness of 150 nm shows a rise time in the range of a few milliseconds while the decay time is slightly longer for pure ZnO film. After doping with Al, the rise time increases to about ten milliseconds and in the same way for the decay time. The results suggest that ZnO based films fabricated by inkjet printing technique are promising and cost efficient solution for developing applications of UV sensors in the future.

ACKNOWLEDGMENTS

The authors acknowledge Dr. Tarja Volotinen for optical measurements, and particularly grateful to Drs. Parmanand Sharma, Valter Ström, Jun Xu and O.D.Jayakumar for valuable discussions. This work is supported by the Swedish Funding Agency VINNOVA, Swedish Research Council, and the Hero-M Excellence Center at KTH. JX is thankful to Carl Tryggers Stiftelsen for a visiting researcher grant. YW is grateful for a graduate study fellowship by the Chinese Scholarship Council -KTH scientific exchange program.

REFERENCES

[1] H.Sirringhaus. Adv. Mater, 17, 2411-2425 (2005).
[2] C. D. Dimitrakopoulos, P. R. L Malenfant. Adv. Mater, 14, 99-117 (2002).
[3] B. A.Ridley, B. Nivi, J. M. Jacobson. Science, 286, 746-749 (1999).
[4] D.C. Look. Mater. Sci. Eng. B, 80, 383-387 (2001).
[5] Y. Takahashi, M. Kanamori, A. Kondoh. Jpn.J.Appl. Phys. 33, 6611-6615 (1994).
[6] P. Sharma, K. Sreenivas, K.V. Rao. J.Appl. Phy, 93, 3963-3970 (2003).
[7] A.E.J.Gonzalez, J.A.S.Urueta, R.S.Parra, J. Crystal Growth, 192, 430-438 (1998).
[8] Z.Q. Xu, H. Deng, J. Xie, Y. Li, X. T. Zu. Appli. Surf. Sci,253, 476-479 (2006).
[9] P.L. Wang, H. Deng, M. Wei. Semic. Optoele.29, 666-668 (2008).
[10] B.Cullity. Elements of X-ray Diffraction, 2nd ed. (Addison-Wesley, 1978), p102.
[11] Yong-Sung Kim, Weon-Pil Tai. Appli. Surf. Sci. 253, 4911-4916 (2007).
[12] A.V.Singh, R. M. Mehra, A.Yoshida and A. Wakahara. J. Apli. Phy. 95, 3640-3643 (2004).
[13] M. Liu, H.K. Kim. J.Appl.Phys.Lett. 84, 173-175 (2004).
[14] O. Dulub, L.A. Boatner and U. Diebold, Surf. Sci.519, 201-2017(2002).
[15] A. Rothschild, Y. Komem and N. Ashkenasy, J. Appl. Phys. 92, 7090-7097 (2002).

Mater. Res. Soc. Symp. Proc. Vol. 1161 © 2009 Materials Research Society   1161-I04-06

Fabrication and Magneto-Capacitance Measurements of PbNb0.02Zr0.2Ti0.8O3/ La0.7Sr0.3MnO3/SiO2/Si Structure Grown by Chemical Solution Deposition

Sushma Kotru and Harshan V Nampoori
Department of Electrical and Computer Engineering, The University of Alabama, Alabama 35486, U.S.A.

ABSTRACT

Multiferroic composite thin films consisting of $PbNb_{0.02}Zr_{0.2}Ti_{0.8}O_3$ (PNZT) and $La_{0.7}$ $Sr_{0.3}$ MnO_3 (LSMO) were deposited on SiO_2/Si substrates. SiO_2 films were deposited by pulsed electron deposition and LSMO and PNZT films were prepared using chemical solution deposition process using a metal organic deposition route. Individual films and the test structure PNZT/LSMO/ SiO_2/Si were characterized using various characterizing techniques. Preliminary results of magnetic field dependent capacitance (magneto-capacitance) on the test structure are reported. A change in capacitance from 18.92 pf to 5.49 pf is observed as frequency changes from 50 KHz to 1 MHz, when no external magnetic field is applied. When a magnetic field of 330 Oe (positive or negative) is applied, the change in magneto-capacitance is appreciable, with a maximum change of 37 % being observed at a frequency of 1 MHz.

INTRODUCTION

In recent years multiferroic/magnetoelectric materials have gained interest among research community due to their scientific significance and potential applications [1]. In multiferroic materials two or three of ferroelectricity, ferromagnetism and ferroelasticity occur in the same phase [2]. However, materials with simultaneous ferroelectric and magnetic ordering are rare in nature [3]. Also magneto electric coupling (α_E) of the single phase compounds is small at room temperature to realize any device applications. Many approaches are being pursued to study the interplay of applied magnetic field and electrical polarization via mechanical coupling between the constituents [4, 5]. Researchers are focusing on artificially engineered multiferroic heterostructures/ composites. In composites, the magneto-electric effect is generated as the product property of a magnetostrictive and piezoelectric compound. An applied magnetic field induces strain in magnetostrictive constituent which gets passed on to piezoelectric constituents, which in turn induces an electrical polarization and vice versa.

Among various ferromagnetic materials, manganites have attracted interest, as one of the potential candidates for magnetoelectric composites due to their high values of magnetostriction [6] and metallic conductivity, which eliminates need of metallic bottom electrode. Among piezoelectric materials, lead zirconate titanate (PZT) is a reasonable choice due to large values for piezoelectric coefficient. For composites, best reported values of coupling constant are 1400 $mVcm^{-1}$ Oe^{-1} for $NiFe_2O_4$/PZT, 4800 $mVcm^{-1}$ Oe^{-1} for Terfenol-D/PZT and 60 $mVcm^{-1}$ Oe^{-1} for $La_{0.7}Sr_{0.3}MnO_3$/PZT multilayer structures [2] . Most of the work though is being carried out on bulk and single crystal substrates. However, to realize devices for practical applications based on these materials, it is essential to integrate such structures with Si.

This paper reports fabrication process of a test structure of $PbNb_{0.02}Zr_{0.2}Ti_{0.8}O_3$/ $La_{0.7}Sr_{0.3}MnO_3$ on Si/SiO_2 substrate along with preliminary results on magneto-capacitance measurements on such structure. Doping PZT with Nb, stabilizes the structure and prevents formation of pyrochlore phase results in single-phase films [7, 8]. Doping also increases the electrical resistance, lowers the coercivity and increases the remanent polarization [8]. We have used PZT (20/80) with 2 % Nb doping for this work.

EXPERIMENT

A test structure of PNZT/LSMO/SiO$_2$/Si was fabricated using the chemical solution deposition method. The SiO$_2$ films were deposited on (100) Si substrates using pulsed electron deposition (PED) technique [9]. In this method an electron gun is used as a source for generating a pulsed electron beam with diameter of about 1 mm. The electron beam acts as a source to generate plasma from a ceramic target of SiO$_2$ (commercially obtained). Distance between source and target was maintained around 3 mm. Films were deposited in oxygen pressure of 4.1 mtorr and substrate temperature was maintained at 650 °C during the deposition. Film thickness was found to be 0.05Å for each pulse. Thus 10,000 pulses were used to obtain 0.5μm thick films of SiO$_2$. La$_{0.7}$Sr$_{0.3}$MnO$_3$ (LSMO) films were deposited on Si/SiO$_2$ using metal organic deposition (MOD) route [10]. Solution for LSMO was prepared in house. To prepare a 0.4 molar solution of LSMO appropriate acetates of La, Sr and Mn were weighed according to the stoichoimetric weight ratios and dissolved in 2-methoxyethanol. LSMO films were deposited using spin coating technique at room temperature. The LSMO solution was deposited on Si/SiO$_2$ substrate using a speed of 4000 rpm for 30 sec. Each layer was heated at 120 °C to get rid of any water followed by a pyrolyzing step at 400 °C to remove the organics. Thickness of each layer was found to be ~ 70 nm. This process was repeated to obtain a thickness of 210 nm. These films were post annealed in tube furnace for 8 hours at 800 °C to achieve better crystallinity [11]. PbNb$_{0.02}$Zr$_{0.2}$Ti$_{0.8}$O$_3$ (PNZT) films were deposited on LSMO/SiO$_2$/Si to complete the heterostructures. PNZT films were also prepared using MOD route [10]. Solution for PNZT was prepared by dissolving corresponding acetates of Pb and Zr in the solvents 2-Methoxyethanol and 2,3 Pentadione respectively . Ti-butoxide and Nb-ethoxide was added to the solution to get the desired stoichiometry. PNZT solution was coated with spin speed of 4000 rpm for 30 sec with each layer being hydrolyzed at 120 °C and pyrolyzed at 400 °C. This process was repeated to obtain the desired thickness of 210 nm. Complete structure (PNZT/LSMO /SiO$_2$/Si) was annealed at 650 °C for 5 minutes with 5000 sccm O$_2$ flow using a rapid thermal annealing process (RTA). Further details about PNZT solution preparation and film deposition can be found in our earlier papers [12, 13].

The heterostructure as well as the individual films of LSMO and PNZT were characterized for phase and orientation using Philips Expert X-ray diffraction (XRD). Magnetization of individual LSMO films as a function of temperature was measured using SQUID (Oxford instruments). Roughness of the films was observed using an AFM from Vecco instruments. The cross section was observed by scanning electron microscopy (FE-SEM) JEOL 2000. For measuring electrical properties, 1000 Å thick platinum electrodes with a diameter of the order of 50-100 μm were used. These electrodes were sputtered on to PNZT layer using a shadow mask and a sputtering unit from Denton Vacuum. LSMO being conducive served as bottom electrode. To expose the LSMO film, PNZT was dry etched using an Intel Vac ion mill. Capacitance measurements were carried out using a probe station along with an LCR meter 4284A from HP. A pair of DC electromagnets powered by a Keithley 228A current source was used to generate the magnetic field. PE curves were obtained from a ferroelectric tester RT 6000 from Radiant technology.

RESULTS AND DISCUSSION

Before fabricating the heterostructure of PNZT/LSMO/SiO$_2$/Si, growth parameters for all individual films were optimized and these films were characterized. The SiO$_2$ films prepared on Si by PED were amorphous. There was no diffraction pattern observed for the films even after annealing the films at 650 °C. RMS roughness as measured by AFM is seen to be 135.6 nm. Before depositing LSMO films on Si/SiO$_2$ substrates, films were deposited on single crystal substrates LAO (LaAlO$_3$) to check quality of the solution prepared as well as that of the grown films. Figure 1(a) shows the XRD patterns of LSMO films grown on SiO$_2$ and LAO. Films grown on LAO were annealing at 800 °C in a tube furnace in flowing oxygen. As expected LSMO films grow epitaxially on single crystal substrates [14, 15] whereas films are polycrystalline when grown on Si/SiO$_2$.

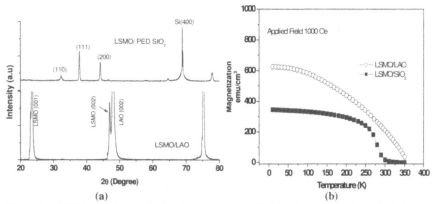

Figure 1. (a) XRD patterns of LSMO grown on SiO$_2$ and LAO substrates, the intensity being plotted on log scale. (b) Magnetization measurement as a function of temperature for LSMO films grown on SiO$_2$/Si and LAO substrate.

Figure 1(b) shows the magnetic moment of films measured in the temperature range of 10K to 350K using a SQUID. The magnetic field applied was 1000 Oe. LSMO films on (001) LAO substrate show a T$_c$ of nearly 360K and saturation magnetization (M$_s$) of about 600 emu/cc. These values are very close to bulk single-crystal films [11, 14], confirming the high quality of sol-gel grown films. LSMO films grown on amorphous SiO$_2$, show lower T$_c$ and M$_s$, however, a significant magnetization is obtained at room-temperature.

To check quality of PNZT films and optimize the growth process, films were initially prepared on Pt/Si substrates. The XRD results from Figure 2(a) show that the film grows with a preferred orientation of (110) on Pt substrates. Figure 2(b) is P-E curve measured at different voltages> the maximum polarization of 80 µC/cm^2 and a coercivity of 100 kV/cm^2 is observed. Finally a test structure PNZT/LSMO/SiO$_2$/Si was prepared. Figure 3(a) shows XRD result on the heterostructure confirming the presence of both phases. It can be seen that PNZT film grown on LSMO shows polycrystalline behavior. An SEM micrograph showing LSMO/SiO$_2$ and PNZT/LSMO interfaces is shown in Figure 3(b).

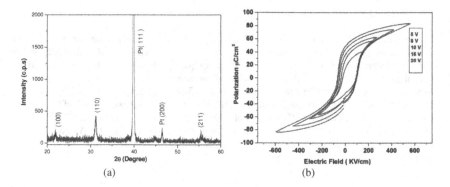

(a)   (b)

Figure 2. (a) XRD pattern of the PNZT films grown on Pt substrates (b) P-E measurements on PNZT films grown on Pt/Si wafers, film thickness is 340 nm.

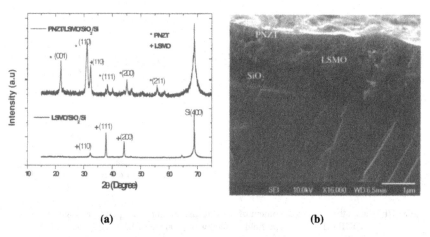

(a)   (b)

Figure 3. (a) XRD pattern on (LSMO/SiO$_2$/Si and heterostructure (PNZT/LSMO/SiO$_2$/Si) (b) Cross section SEM image of PNZT/LSMO/SiO$_2$ showing interface of the heterostructure.

Magneto-capacitance (MC) change was calculated using the equation 1 [16, 17]

$$MC\% = \left[\frac{C_p(0) - C_p(H)}{C_p(H)} \right] \times 100 \qquad (1)$$

142

where $C_p(0)$ and $C_p(H)$ corresponds to capacitance measured without magnetic field and with the field (H). Figure 4 shows capacitance and change in magneto-capacitance of the structure as a function of frequency. A DC magnetic field of 330 Oe was applied in positive and negative direction. This field was chosen from the M-H curve measured by VSM on the LSMO films not reported here (coercivity ~ 290 Oe). A change in capacitance from 18.92 pf to 5.49 pf is observed as the frequency changes from 50 KHz to 1 MHz, when no external magnetic field is applied. When a magnetic field of 330 Oe (positive or negative) is applied, the capacitance changes from 18.93 pf to 3.96 pf as the frequency changes from 50 KHz to 1 MHz. It is observed that capacitance of the heterostructure decreases when the field is applied; the trend being same when the magnetic field is applied in either positive or negative direction. Change in capacitance with and without the magnetic field (magneto-capacitance) is smaller at lower frequencies and much appreciable at higher frequencies, with the maximum change of 37 % in magneto-capacitance being observed at a frequency of 1 MHz.

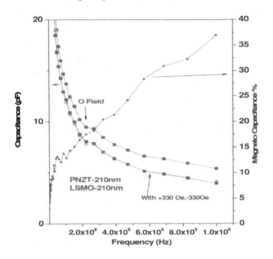

Figure 4. Capacitance and magneto-capacitance measured on the PNZT/LSMO/SiO$_2$/Si heterostructure as a function of frequency, the applied magnetic field is ± 330 Oe

CONCLUSIONS

We have successfully fabricated a test structure of LSMO/PNZT on Si/SiO$_2$ by chemical solution deposition method. Preliminary results show that external magnetic field changes the capacitance of structure which suggests the possibility of intrinsic magneto electric response from such structure. The percentage change in magneto-capacitance was seen to increase with frequency, the maximum change being 37 % at a frequency of 1 MHz. Further measurements are in progress.

ACKNOWLEDGMENT

One of the authors acknowledges the financial support from ECE department and MINT center. Authors also extend their thanks to the MINT center for use of characterization facilities. This work used Central Analytical Facility, which is supported by the University of Alabama.

REFERENCES
[1] J. F. Scott, Nature Materials, **6** (2007) 256-257.
[2] W. Eerenstein, N. D. Mathur, J. F. Scott, Nature, **442** (2006) 759-765.
[3] N. A. Hill, Journal of Physical Chemistry B, **104** (2000) 6694-6709.
[4] M. Fiebig, Journal of Physics D-Applied Physics, **38** (2005) R123-R152.
[5] R. Ramesh, N. A. Spaldin, Nature Materials, **6** (2007) 21-29.
[6] G. Srinivasan, E. T. Rasmussen, B. J. Levin, R. Hayes, Physical Review B, **65** (2002).
[7] K. W. Kwok, R. C. W. Tsang, H. L. W. Chan, C. L. Choy, Journal of Applied Physics, **95** (2004) 1372-1376.
[8] T. Haccart, E. Cattan, D. Remiens, S. Hiboux, P. Muralt, Applied Physics Letters, **76** (2000) 3292-3294.
[9] R. J. Choudhary, S. B. Ogale, S. R. Shinde, V. N. Kulkarni, T. Venkatesan, K. S. Harshavardhan, M. Strikovski, B. Hannoyer, Applied Physics Letters **84** (2004) 1483-1485.
[10] F. F. Lange, Science, **273** (1996) 903-909.
[11] A. Urushibara, Y. Moritomo, T. Arima, A. Asamitsu, G. Kido, Y. Tokura, Physical Review, **B** 51 (1995) 14103.
[12] H. Han, X. Y. Song, J. Zhong, S. Kotru, P. Padmini, R. K. Pandey, Applied Physics Letters, **85** (2004) 5310-5
[13] H. Han, S. Kotru, H. Zhong, R. K. Pandey, Infrared Physics & Technology, **51** (2008) 216-220.
[14] S. M. Liu, X. B. Zhu, J. Yang, W. H. Song, J. M. Dai, Y. P. Sun, Ceramics International, **32** (2006) 157-162.
[15] A. Huang, K. Yao, J. Wang, Thin Solid Films, **516** (2008) 5057-5061.
[16] J. X. Zhang, J. Y. Dai, W. Lu, H. L. W. Chan, B. Wu, D. X. Li, Journal of Physics D: Applied Physics, **41** (2008) 235405.
[17] A. R. Chaudhuri, P. Mandal, S. B. Krupanidhi, A. Sundaresan, Solid State Communications,**148** (2008) 566-569.

Mater. Res. Soc. Symp. Proc. Vol. 1161 © 2009 Materials Research Society 1161-I05-03

Flexible Ceramic-Polymer Composite Substrates With Spatially Variable Dielectrics for Miniaturized RF Applications

Zuhal Tasdemir[1] and Gullu Kiziltas[2]

[1]Materials Science and Engineering, Sabanci University, Orhanli, 34956 Istanbul, Turkey
[2]Mechatronics Engineering, Sabanci University, Orhanli, 34956 Istanbul, Turkey

ABSTRACT

The goal of this research is to develop a process suitable for producing monolithic conformal substrates with a spatial arrangement of material cells according to a particular design creating novel material systems, useful for many multi- functional electronic and Radio Frequency devices. In this study, MCT ceramics (Mg-Ca-Ti-O systems) and organic binders (polymer solution) are mixed and fabricated as films through a process called tape casting to compromise between high dielectric constant and flexibility. Prior to optimizing the process, several characterization studies are carried out: Commercial spray dried MCT powders (Transtech Inc.) with dielectric constant k=70 and k=20 were analyzed as pressed and produced into tape cast films. Dielectric properties are then measured by an Agilent 16451B material analysis kit and their microscopic behavior is examined by scanning electron microscopy. Results show that flexible composite films show a maximum dielectric constant of $\varepsilon \sim 22$ unlike their powder pressed form with $\varepsilon \sim 16$ but their loss behavior deteriorates when compared with their sintered form and a loss tangent factor of 0.001. The difference is attributed to the air content vs. polymer presence of the material in powder pressed form. Also, these substrates naturally are no longer flexible; hence studies are focused on their tape cast form. The potential of these dielectric shades to serve as candidate constituents for producing monolithic textured polymer-ceramic-composites with controllable loss is studied further. Four properties are of prime importance: tunability of dielectric constants to achieve miniaturization, flexibility via low temperature processing of polymers and loss controllability.

INTRODUCTION

Microwave dielectric materials play an important role in achieving devices with enhanced performance for a wide range of Radio Frequency applications [1]. In order to meet the stringent needs of these systems such as deformability, low-loss and miniaturization, improved or novel microwave components based on dielectric materials and their new designs are required. Dielectric materials can be either organic as in the example of polymers or inorganic like ceramics. With its wave guiding ability the choice of material directly impacts device performance such as propagation speed (hence miniaturization) and the characteristic impedance (hence mismatch losses). In this study, we try to achieve a high dielectric constant by producing composite materials composed of both ceramics and polymer in order to achieve both miniaturization and deformability such as in [2]. The goal here is to analyze their potential in producing textured substrates for future performance enhancements as was suggested earlier via ceramic constituents only [3]. A high dielectric constant of the final composite is achieved by using ceramic materials, specifically Mg-Ca-TiO$_2$ (MCT ceramics) are used in this study. These ceramics are mixed with polymers to obtain desired deformability of the resulting composite substrate. In the present study, ceramic-polymer composites films are produced via tape casting process. These films are characterized by measuring their dielectric behavior and their microstructures are investigated through Scanning Electron Microscopy (SEM). Towards the final goal of achieving monolithic conformal substrates with a dielectric spatial distribution, warm binding is used for the assembly of square pixels made of MCT green tapes. To test the resulting conformal substrate, a simple patch antenna is constructed using the resulting mosaic substrate. To demonstrate the feasibility of the proposed approach a miniaturized patch antenna is fabricated via tape cast film substrates with 2D material variation.

EXPERIMENTAL PROCEDURE

Before producing and characterizing polymer-ceramic composite tape cast films, pellets of constituent ceramic powders were made of k=70 MCT (MCT-70) and characterized to assess their dielectric behavior. To analyze MCT's dielectric tunability and loss controllability, pellets were produced by three different processes; 1) Uniaxially pressing them at 75 MPa using available spray-dried ceramic powders, 2) Sintering the pellets produced at step 1 at 550°C for 1 hr, and 3) Pre-sintering the powder itself at the burnout temperature 550°C for 1hr and then pressing them into pellets at 75 MPa. Having knowledge on the dielectric behavior limitation of the constituent ceramic powder and the effect of pre-sintering on the powder pressed pellets, similar analysis was made for tape cast films. The casting slurry based on MCT was prepared first. Specifically, the chosen MCT powder and the polymer solution composed of organic solvents, binder and the dispersant were mixed at 50% weight ratio and ball-milled for 12 h to obtain a uniform mixture (Figure 1 a). After degassing the slurry in the magnetic stirrer for 5 hours, the slurry was cast on the glass layer to form a 130 μ- thick layer using a doctor blade (Figure 1.b). Multiple cast films (Figure 1 c) are then stacked onto each other, as shown in (Figure 1 d) and pressed uniaxially at 90 MPa to obtain a laminar structure shown in Figure 1 e.

a. Ball milling process   b. Tape casting   c. Laminar films

d. Stacked film layers before pressing   e. Stacked layers after pressing
Figure 1.a-e Tape casting process

TGA analysis of films were carried out to determine the temperature to burn off the polymer which was determined as 550°C. Laminated films are characterized next. These films were made using four different powders: 1) As-is powders, 2) pre-sintered powders at 550°C and 3) pre-sintered powders at 1360°C which were either ball-milled (uniform) or pounded in a mortar (non-uniform). Also, the effect of using pre-sintered powders for tape casting at 550 °C and 1350 °C is analyzed. The powder sintered at 1360 °C is processed in two different ways using ball milling and pounding in a mortar. For the dielectric measurements, pellet samples are prepared and measured between 40 Hz- 30 MHz using the Agilent 16451B Material Analysis kit with the 4294A network analyzer. Their microstructures are analyzed using SEM. Finally, a simple mosaic substrate (Figure 2) is produced via machining and assembling films produced via tape casting of of two shades of MCT ceramic (k=20 and k-20) mixed with polymer. Resulting layout is uniaxially pressed to form the conformal substrate of a patch antenna to be tested for miniaturization purposes shown in flat and deformed configurations in Figure 2.

Figure 2. Mosaic substrates made of tape cast films using k=20 (white) and k=70 (brown) ceramic-polymer mixtures shown in flat (left) and deformed (right) configuration.

DISCUSSION

Dielectric measurements for pellets made of MCT 70 and three different processes explained above are shown in Figure 3a. It is observed that the highest dielectric constant, ε~ 16.5 corresponds to pellets made from available spray-dried powder. This is because 550°C does not correspond to the full dense sintering temperature of the MCT ceramic pellets, i.e. they are only partially dense, hence their porous microstructures result in lower dielectric constants. Nevertheless, powders were not fully sintered to 1360°C and heat treatments at higher temperatures were avoided in order to maintain flexibility of resulting films. Also, post-sintered ceramic pellets indicate a slightly higher dielectric constant than pellets produced via pre-sintered ceramic powders due to higher density. In Figure 3b, dielectric loss tangent measurements of the same pellets are given. The lowest possible loss value around ~ 0.0025 is achieved for sintered ceramic pellets. This is because the amount of pores decreases during the firing process, and therefore loss value decreases. Also, loss behavior of ceramic pellets improves slightly via pre-sintering of powders from 0.016 to 0.012. As shown in Figure 4a, dielectric constant value reaches ~24 for films made from spray dried ceramic powder and decreases to ~17 for films made from powder pre-sintered at 1360°C. Also, pre-sintering temperature and powder processing have a tuning effect on the resulting dielectric constant of the composite film as has been observed by the increasing effect of dielectric constant for non-uniform powders produced after pre-sintering.

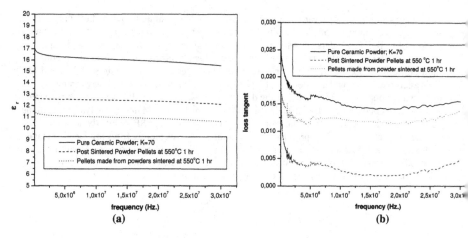

Figure 3. (a) Dielectric constant and (b) loss tangent vs. frequency (Hz) of MCT70 powder pressed pellets produced via three different processes.

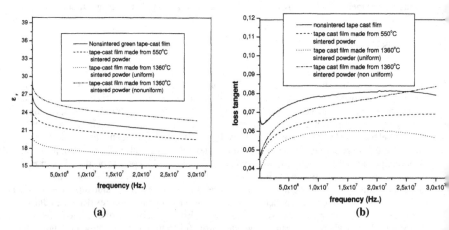

Figure 4. (a) Dielectric constant and (b) loss tangent vs. frequency (Hz) of tape cast films made of MCT70 ceramic powders via four different processes.

As for the loss tangents of tape cast films in Figure 4b, it is seen that a film with a high dielectric constant (Figure 4a) is accompanied with a high loss tangent value. The lowest loss tangent value is observed for films made from 1360°C sintered powders which were ball-milled prior to tape

148

casting. An unusual trend is observed for non-uniform powder films which are pre-sintered at high temperatures. Therefore, the effect of pre-sintering and powder processing prior to tape casting needs further investigation. As sintering of MCT's is expected to reduce the grain size and produce a more uniform microstructure, loss value is expected to decrease, but the same reduction was not observed for the dielectric constant value. Based on comparison of powder pellets and tape cast film behavior in Figure 3 and Figure 4, respectively, there is an overall increase of loss tangent value by a factor of 10 for tape cast films. This indicates that other mechanisms for further loss control could be investigated.

Microstructures of spray dried ceramic powders are analyzed via SEM and powder particles were observed to have a distribution as shown in Figure 5 with their particle size ranging from 20 to 180 μm for k= 20 (a) and k=70 (b).

(a)   (b)

Figure 5. SEM image of **a)** MCT70 and **b)** MCT20 ceramic powder.

SEM revealed good adhesion between the polymer and ceramic particles for tape cast green films (Figure 6a). Although some ceramic agglomerations are observed, there is an overall uniform distribution in the polymer matrix phase. When the same sample is sintered at 550°C for 1 hr, polymer evaporates mostly leaving ceramic agglomerates as the primary phase in the mixture (Figure 6b).

(a)   (b)

Figure 6. SEM image of ceramic/polymer composite **(a)** in green state and **(b)** after sintering at 550°C for 1 hr.

149

After producing the mosaic structure, copper tape is mounted on the substrate to produce a radiating patch as shown in Figure 8. Patch antenna will be fed by a coaxial cable and measurement results will be presented at the conference.

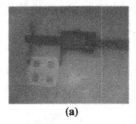 

(a)   (b)

Figure 8. Mosaic green tape structure in (a) substrate form and (b) with copper tape (patch) and probe

CONCLUSIONS

In this study, the dielectric behavior of laminated tape cast composite films made of MCT based ceramic powder and polymers are investigated. In addition, the effect of firing tape cast films at the burnout temperature of the polymer as well as producing tape cast films using pre-sintered ceramic powders at burn-out and full-sintering temperature is analyzed. It is found that firing temperature and the post processing of powders prior to tape casting have direct effect on both the dielectric constant and loss of resulting films. Results prove tunability and loss controlling capability of resulting deformable substrates. More detailed analysis of the relationship between microstructure and dielectric behavior is currently underway. Finally, MCT green tape films were used to produce a 2D textured conformal substrate for conformal antenna applications.

ACKNOWLEDGMENTS

This work was supported by the career grant # 106M348 supported by The Scientific and Technological Research Council of Turkey and TUBA GEBIP Award 2008.

REFERENCES

[1] J. L.Volakis, G. Mumcu, K. Sertel, C.-C. Chen, M. Lee, B. Kramer, D. Psychoudakis, G. Kiziltas, IEEE Antennas and Propagation Magazine **48** (5), 12-28 (2006).
[2] S. Koulouridis, G. Kiziltas, Y. Zhou, D. Hansford, J. L. Volakis, IEEE Transactions on Microwave Theory and Techniques **54**, 4202-4208 (2006).
[3] Z. N. Wing and J. W. Halloran, J. Am. Ceram. Soc., **89** (11), 3406–12 (2006).

Mater. Res. Soc. Symp. Proc. Vol. 1161 © 2009 Materials Research Society   1161-I05-04

Emerging Technologies and Opportunities Based on the Magneto-Electric Effect in Multiferroic Composites

Marian Vopsaroiu, John Blackburn, Markys G. Cain
National Physical Laboratory, Teddington TW11 0LW, United Kingdom

ABSTRACT

Multiferroic materials are recognized today as one of the new emerging technologies with huge potential for both academic research and commercial developments. Multiferroic composites are in particular more attractive for studies due to their enhanced properties, especially at room temperature, in comparison to the single-phase multiferroics. In this paper, we examine some of the theoretical aspects regarding one type of multiferroic composites (laminated structures) and we discuss one of the many possible applications of these exciting structures. We highlight the main advantages composite systems have over single-phase multiferroics and the similarities that exist between them.

INTRODUCTION

Continued advances in high technologies, electronics and computing are possible due to the continue miniaturization of solid-state electronics. This is achieved by reducing the physical size of the active element (i.e. memory element, logic gate, transistor, etc) or by engineering new materials with multifunctional features capable of performing more than one operation / function in the same physical volume. Multiferroic (MF) materials are excellent candidates for new technologies and applications because they exhibit multiple cooperative phenomena (i.e. magnetic ordering, electric ordering and piezo effects or ferro-elasticity). Moreover, these intrinsic cooperative phenomena can be coupled to each other via the magneto-electric (ME) effect or via the strain mediated (STME) effect, opening up the possibility to create voltage control of magnetization or electric polarization tuning via an external applied magnetic field. So far, a number of important applications based on the ME effect have been proposed or developed, mostly in bulk MF composite materials. Some of the most important applications are: ac magnetic field sensors [1], DC magnetic field sensors [2], current sensors [3], transformers and gyrators [4], microwave devices [5], FMR resonators and filters [6], electric field tuneable FMR phase shifters [7] and hybrid spintronic - MF devices as potential MRAM type of memory element [8]. In this paper we focused our attention on the advantages that composite MFs have against single phase MFs. We also discuss the implications of the strain mediated ME coupling in MF composites and we introduce some theoretical aspects of the composite MFs, with emphasis on a specific structure: a tri-layer laminated MF bar. This type of thin film MF device is then applied to the design of a magnetic recording read head sensor presented in the final section of this paper. A direct comparison between current magneto-resistive technologies used in magnetic recording read heads and the new proposed MF read head technology is given. Theoretical calculations of such a MF recording read head show that the technology can meet the requirements for 1 Tb/in^2 magnetic recording densities. This capability together with a number of considerable advantages could in future motivate the replacement of conventional magneto-resistive (MR) read heads with MF ones.

COMPOSITE VS SINGLE PHASE MULTIFERROICS

Although MF materials have been known for a long time as single-phase compounds [9-12], in recent years there has been an increased interest in MF due to the realization of their potential in composite form [13,14].

Table 1. Single-phase multiferroics. The value of the ME coupling is given for measurements at $T < RT$. A missing value means that there are no reported measurements for the specific sample.

MF Material	Transition temp. (K)	ME coupling coefficient	Ref
Cr_2O_3	307	0.74 mV/cm Oe	15, 16
$Cr_3B_7O_{13}Cl$	9.7, 13.5	-	16
$GaFeO_3$	350	-	17
$LiCoPO_4$		5.52 mV/cm Oe	16, 18
$TbPO_4$	$T_n = 2.27$	6.62 mV/cm Oe	19
$Pb(Fe_{1/2}Ta_{1/2})O_3$	$T_n = 150$	-	56
YIG	-	5.41 mV/cm Oe	20
$BiFeO_3$	$T_c^e = 1103$, $T_n = 643$	-	21
$BiMnO_3$	$T_c^e = 450$, $T_n = 100$	-	21
$CdCr_2S_4$	$T_c^e = 135$	-	22
$YMnO_3$	$T_c^e = 640$; $T_n = 100$	-	23, 21
$HoMnO_3$	$T_c^e = 875$; $T_n^{Mn} = 75$; $T_n^{Ho} = 4.6$	-	24
$TbMnO_3$	$T_c^e = 27$; $T_n^{Mn} = 42$; $T_N^{Tb} = 7$	-	25
$TbMn_2O_5$	$T_c^e = 38$; $T_n^{Mn} = 43$; $T_n^{Tb} = 10$	-	26
$BaMnF_4$	$T_n = 26$	-	27
$Ba_{0.5}Sr_{1.5}Zn_2Fe_{12}O_{22}$	$T_n = 326$	-	28
$PbFeTiO_3$	$T^m = 270$; $T_c^e = 420$	-	29

Table 2. Multi-phase multiferroics. (Terfenol D = $Tb_{1-x}Dy_xFe_2$; PZT = $PbZr_{1-x}Ti_xO_3$). The indicated transition temperatures are for the magnetic phase, while $T_c^{PZT} \approx 650$ K. A missing value means that there are no reported measurements for the specific compound.

MF Material	Structure	Transition temp. (K)	ME coupling coefficient	Ref
$CoFe_2O_4$ in $BaTiO_3$ matrix	Composite	T >RT	50 mV/cm Oe	30
$BaTiO_3$-$CoFe_2O_4$	Nano-structured composite	$T_c^E = 390$; $T_c^m > RT$	-	55
$Tb_{1-x}Dy_xFe_2$ / PZT in polymer matrix	Composite	T >RT	3000 mV/cm Oe	31
$Tb_{1-x}Dy_xFe_2$ / PZT PZT / $Tb_{1-x}Dy_xFe_2$ / PZT	Laminated composite	T > RT	4800 mV/cm Oe	32
Terfenol D/ $Pb(MgNb)O_3$- $PbTiO_3$	Laminated composite	T >RT	2200 mV / cm Oe	58
$La_{0.7}Sr_{0.3}MnO_3$ / PZT	Laminated composite	$T_c^m = 260$	60 mV/cm Oe	33
$La_{0.7}Sr_{0.3}MnO_3$ / $BaTiO_3$	Bi-layers	T = 278 lattice change	-	34
CoPd / PZT	Bi-layers	-	-	35
$NiFe_2O_4$/PZT	Particulate composite	T > RT	80 mV/cm Oe	57
$NiFe_2O_4$ / PZT	Laminated comp	T > RT	1400 mV/cm Oe	36, 37
CoFeV / PZT / CoFeV	Tri-layers	$T_c^m = 1213$	90000 mV / cm Oe	38
$Ni_{0.8}Zn_{0.2}Fe_2O_4$ / PZT	Bi-layers	T > RT	600 mV / cm Oe	39
$Co_{0.6}Zn_{0.4}Mn_2O_4$ / PZT	Bi-layers	T > RT	250 mV / cm Oe	39

This is because the ME coupling in single phase MF is very weak and usually occurs at low temperatures. In the case of MF composite, the material can be engineered so that the individual phases are optimized for room temperature operation and enhanced ME coupling, which in turn is translated in more sensitive sensors. Moreover, the composite MF could be used in bulk as well as thin film structures, which makes them the best candidates for vertical integration at wafer level. This is well emphasised in the Tables 1 and 2, where some of the published single phase and composite MFs are listed together with their transition temperature and magnetically induced coupling strength, where values are available. ME coupling coefficients clearly indicate that multi-phase compounds (Table 2) exhibit at least one order of magnitude larger couplings than the single-phase compounds (Table 1).

MAGNETO-ELECTRIC COUPLING IN COMPOSITES AND SINGLE-PHASE MUTIFERROICS

The interplay between electric and magnetic states is realized via the (ME) effect, which is defined as the coupling between electric and magnetic fields in matter. Although this coupling can have non-linear components, the ME effect is usually described mathematically by the linear ME coupling coefficient (α), which is the dominant coupling term [40,41]. In multiferroics, the internal magnetic / electric fields are enhanced by the presence of the multiple long-range ferroic ordering, which in turn produces large ME coupling effects. In addition, the ME coupling can be stress mediated in samples exhibiting piezo-effects, especially when the two ferroic phases are clearly separated as in MF laminates or composites. In this case, a device effective ME coupling coefficient α^{eff}, which contains the linear direct ME effect and the stress-mediated component, describes the ME effect [42].

Assume that a MF material contains both electric (E) and magnetic (M) phases. In addition, assume that the two ordered phases exhibit piezo effects, piezo coupling effects and ME coupling. The coupling between thermal, electric, magnetic and elastic parameters of the material can be expressed using a thermodynamic approach. This will provide relations between macroscopic material parameters, which can be measured in different experimental conditions. Since in most situations we work at constant temperature T and use the external applied fields and stress as independent variables, it is useful to use the Gibbs free energy G(T, σ, H, E) as the thermodynamic potential, which for this MF system is expressed as:

$$-G(T, \sigma, E, H) = ST + \frac{1}{2} s_{ijkl} \sigma_{kl} \sigma_{ij} + d^{e}_{ijk} \sigma_{ij} E_k + d^{m}_{ijk} \sigma_{ij} H_k + \pi_{ijkl} E_i H_j \sigma_{kl}$$

$$+ \frac{1}{2} \varepsilon_{ij} E_i E_j + \frac{1}{2} \mu_{ij} H_i H_j + \alpha_{ij} E_i H_j + \dots \tag{1}$$

where: S is the entropy of the system; s_{ijkl} is the elastic compliance fourth rank tensor; σ_{kl} is the mechanical external applied stress tensor; d^{e}_{kij} is the third rank tensor of piezoelectric coefficient; d^{m}_{kij} is the third rank tensor of piezomagnetic coefficient; π_{ijkl} is the piezo-coupling constant fourth rank tensor; ε_{ij} is the dielectric permitivity tensor; μ_{ij} is the magnetic permeability tens or of the material; α_{ij} the ME linear coupling coefficient. The $\alpha_{ij} E_i H_j$ represents the ME energy term, while higher order non-linear coupling terms are not considered here. From (1) we can derive the total strain (x) of the system:

153

$$x_{ij} = -\left(\frac{\partial G}{\partial \sigma_{ij}}\right)_{T,E,H} = s_{ijkl}\sigma_{kl} + d^e{}_{ijk}E_k + d^m{}_{ijk}H_k + \pi_{ijkl}E_kH_l \tag{2}$$

where: the first term represent the strain due to the external applied stress, the second term is the piezoelectric strain due to the application of the electric field E, the third term is the piezomagnetic strain due to the applied magnetic field H and the final term is a piezo-coupling strain induced by the mutual piezoelectric / piezomagnetic effects in the sample (i.e. strain mediated ME effect). Performing standard thermodynamic differentiation of (1), the general expression for the electric polarization and magnetization of a multiferroic system leads to:

$$P_i = -\left(\frac{\partial G}{\partial E_i}\right)_{\sigma,T,H} = \varepsilon_{ij}E_j + d_{ijk}{}^e\sigma_{jk} + \alpha_{ij}H_j + \pi_{ijkl}H_j\sigma_{kl} \tag{3}$$

$$M_i = -\left(\frac{\partial G}{\partial H_i}\right)_{\sigma,T,E} = \mu_{ij}H_j + d^m_{ijk}\sigma_{jk} + \alpha_{ij}E_j + \pi_{ijkl}E_j\sigma_{kl} \tag{4}$$

which can be re-written using condensed matrix notation as:

$$P_i = \varepsilon_{ij}E_j + d_{im}{}^e\sigma_m + \alpha^{eff}{}_{ij}H_j \tag{5}$$

$$M_i = \mu_{ij}H_j + d^m_{in}\sigma_n + \alpha^{eff}{}_{ij}E_j \tag{6}$$

where: i, j = 1, 2, 3 and m, n = 1, 2, 3, 4, 5, 6 and

$$\alpha^{eff}{}_{ij} = \alpha_{ij} + \pi_{ijkl}\sigma_{kl} \tag{7}$$

where α^{eff} is the device effective ME coupling coefficient, which includes the linear ME coupling effect (α) and the piezo-coupling (or stress mediated) ME effect ($\pi\sigma$). In the case that there is no piezo-coupling in the system (i.e. $\pi = 0$) or no piezo effects, then the effective ME coupling coefficient is identical to the linear ME coefficient ($\alpha^{eff} = \alpha$). In this case, however, the equations (5) and (6) are written identically except the α^{eff} is replaced by α. Therefore, using relations (5) and (6), the electrically (8) and magnetically (9) induced ME couplings are expressed identically in the case of direct ME coupling, strain mediated ME coupling or indeed in the case of a ME coupling which involves a combination of both.

$$\alpha_E = \frac{\partial M}{\partial E} \tag{8}$$

$$\alpha_H = \frac{\partial P}{\partial H} \tag{9}$$

where we dropped the matrix indexes. The implications of this are: i) direct or strain mediated ME coupling coefficients are mathematically described identically; ii) a MF device or sensor would physically operate in the same way regardless of the type of ME coupling displayed; iii) experimental measurements of the ME coupling coefficient can be designed identically for any type of ME coupling.

A SIMPLE TRI-LAYER MULTIFERROIC STRUCTURE

In the case of MF laminates, the ME coupling coefficient can be derived analytically using a classical electro-mechanical approach [43]. However, in solving the electromechanical equations of the piezo-MF structures, the boundary conditions should be carefully considered since they dictate the choice of independent variables for the constitutive piezo-equations. Mechanical boundary conditions have usually either free or clamped stress – strain, while the electric and magnetic boundary conditions can be imposed by the geometry, the location of the electroded surfaces and the direction of the applied external magnetic fields. Therefore, the adequate choice of the independent variables and consequently the correct formulation of the constitutive equations can only be made for specific geometries, which are usually "thickness expander plate" and "length expander bar". These issues have not been always carefully considered in the existing literature reporting theoretical derivation of the ME effect in laminated multiferroics. Depending on the choice of the independent variables, there are four different sets of constitutive equations for isothermal processes, which can be written in general form using matrix notations (see Table 3). Throughout the paper we use the superscript "e" for the coefficients of the E layers and the superscript "m" for the M layers, but we drop e and m superscripts where obvious (e.g. $B_1^m = B_1$, etc.). The numbers indicate the matrix components.

Table 3. The general constitutive equations corresponding to piezoelectric and piezomagnetic material for different sets of independent variables and isothermal processes, where: x is the strain, σ is the stress, ε is the dielectric permittivity, μ is the magnetic permeability, s is the elastic compliance, c is the elastic stiffness, d, e, g and h are the piezo-coefficients and β is the inverse of the electric or magnetic susceptibility.

Piezoelectric		Piezomagnetic	
Independent variables	Constitutive equations	Independent variables	Constitutive equations
σ ; E	$x_m^e = s_{mn}^e \sigma_n^e + d_{im}^e E_i$ $D_i = d_{im}^e \sigma_m^e + \varepsilon_{ij} E_j$	σ ; H	$x_m^m = s_{mn}^m \sigma_n^m + d_{im}^m H_i$ $B_i = d_{im}^m \sigma_m^m + \mu_{ij} H_j$
x ; E	$\sigma_m = c_{mn}^e x_n^e - e_{im}^e E_i$ $D_i = e_{im}^e x_m^e + \varepsilon_{ij} E_j$	x ; H	$\sigma_m^m = c_{mn}^m x_n^m - e_{im}^m H_i$ $B_i = e_{im}^m x_m^m + \mu_{ij} H_j$
x ; D	$\sigma_m^e = c_{mn}^e x_n^e - h_{im}^e D_i$ $E_i = - h_{im}^e x_m^e + \beta_{ij}^e D_j$	x ; B	$\sigma_m^m = c_{mn}^m x_n^m - h_{im}^m B_i$ $H_i = - h_{im}^m x_m^m + \beta_{ij}^m B_j$
σ ; D	$x_m^e = s_{mn}^e \sigma_n^e + g_{im}^e D_i$ $E_i = -g_{im}^e \sigma_m^e + \beta_{ij}^e D_j$	σ ; B	$x_m^m = s_{mn}^m \sigma_n^m + g_{im}^m B_i$ $H_i = -g_{im}^m \sigma_m^m + \beta_{ij}^m B_j$

We limit our analysis to the magnetically-induced ME effect in laminated tri-layer multiferroic structures of the type ferromagnetic (FM) / ferroelectric (FE) / ferromagnetic (FM). The MF structure has a bar shape with the length L, width w and total thickness $t = 2t_m + t_e$ (L >> t; w). We also assume isothermal processes. The polarization of the piezo-electric layer is in the OZ direction (i.e. out-of-plane) and the magnetic layers are magnetized in the OX direction (i.e. in-plane) by the application of the magnetic fields along the X direction (see Figure 1). The operation geometry in which the polarization direction of the FE layer is transverse to the magnetic field directions (e.g. the magnetization of the FM layers) is know as transverse geometry and has been proven experimentally to give the largest ME effect [38]. The FE layer

155

has thickness t_e and the two magnetostrictive FM layers, which also act as electrodes, are t_m thick. In general, elastic displacement, ξ, will vary through the thickness of the 3 layers as well as along the length of the bar. Displacement cannot be constant in z because the three layers have different properties and consequently different sound-propagating wavelengths. However, we will calculate ξ as the average displacement of the three layers, ignoring the details of displacements in each layer.

A)

B)

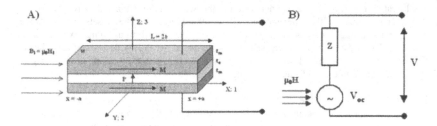

Figure 1. A) Tri-layer multiferroic length extension bar with P and M transverse to each other. B) The equivalent electrical circuit of the device, which operates as a voltage source

Such an approach is best calculated using the surface integral formulation of Newton's Laws:

$$\oint_S \vec{\sigma} \cdot \vec{n} dS = M \frac{d^2 \xi}{dt^2} \tag{10}$$

where M is the mass of the material within the surface S, ξ is the average displacement and σ is the stress on the surface. We assume that the only non-zero variables are $E_z(x) = \text{const}$, $D_z(x)$, $H_x(x)$, $B_x(x) = \text{const}$. The only non-zero stress and strain are $\sigma_1(x)$ and $x_1(x)$ respectively with all other stress components vanishing. With this simplification, applying (10) to the elementary surface gives:

$$\left(2\rho_m t_m + \rho_e t_e\right)\frac{\partial^2 \xi}{\partial t^2} = 2t_m \frac{\partial \sigma_1{}^m}{\partial x} + t_e \frac{\partial \sigma_1{}^e}{\partial x} \tag{11}$$

where ρ_m and ρ_e are the densities of the FM and FE materials, respectively. It is convenient to choose constitutive laws with stress, E-field and B-field as independent variables ((σ; E) and (σ; B) in Table 3). Taking account of ∞mm crystal symmetry along 3 for the FE (e.g. PZT operated in a '31' type mode) or along 1 for the FM (e.g. Terfenol) gives:

$$x_1{}^e = s_{11}{}^e \sigma_1{}^e + d_{31}{}^e E_3 \tag{12}$$
$$D_3 = d_{31}{}^e \sigma_1{}^e + \varepsilon_{33} E_3 \tag{13}$$
$$x_1{}^m = s_{11}{}^m \sigma_1{}^m + g_{11}{}^m B_1 \tag{14}$$
$$H_1 = -g_{11}{}^m \sigma_1{}^m + \beta_{11} B_1 \tag{15}$$

where d, g, s, ε and β are the piezoelectric constant, magnetostrictive constant, elastic compliance (at constant E for the FE and constant B for FM), permittivity and the inverse permeability both at constant stress. Using in (11) – (15) the average displacement approximation ($\xi^e{}_1 = \xi^m{}_1 = \xi_1$), the average strain approximation ($x_1{}^m = x_1{}^e = x_1$) and assuming ac time variation so that $d/dt \rightarrow i\omega t$ throughout, the electric current flowing out of the device is derived as:

$$I_3 = \int_{-a}^{+a} i\omega D_3(x) w dx$$

$$= i\omega w L \left[\left(\varepsilon_{33} - \frac{\left(d_{31}^e\right)^2}{s_{11}^e} + \frac{\left(d_{31}^e\right)^2 s_{11}^m \eta}{s_{11}^e \left(\eta s_{11}^m + (1-\eta)s_{11}^e\right)} \frac{\tan ka}{ka} \right) E_3 + B_1 \frac{d_{31}^e g_{11}^m (1-\eta)}{\left(\eta s_{11}^m + (1-\eta)s_{11}^e\right)} \frac{\tan ka}{ka} \right]$$

(16)

where:

$$k = \omega \sqrt{\frac{(\rho_m(1-\eta) + \rho_e \eta)}{s_{11}^e (1-\eta) + s_{11}^m \eta} s_{11}^e s_{11}^m}$$

(17)

and $\eta = t_e / t$ is the thickness ratio (i.e. $t_e = \eta t$ and $2t_m = t(1-\eta)$)

Assuming that FM layers are highly conductive, then the electric field is concentrated in the piezoelectric region, so that we can write $V_3 = V = -E_3 t_e = -E_3 t \eta$. The device works in fact as a voltage source by converting the magnetic energy (inflicted by the applications of external magnetic fields) into a voltage. Using (16), the open circuit induced voltage V_{oc} can be estimated for the open circuit condition ($I_3 = 0$) as:

$$V_{oc} = \frac{d_{31}^e g_{11}^m s_{11}^e (1-\eta)\eta}{\varepsilon_{33} s_{11}^e \left[\eta s_{11}^m + (1-\eta)s_{11}^e\right] - \left(d_{31}^e\right)^2 \left[(1-\eta)s_{11}^e + \eta s_{11}^m \left(1 - \frac{\tan ka}{ka}\right)\right]} \cdot B_1 \cdot t \cdot \frac{\tan ka}{ka}$$

(18)

where: $V_{oc} = \alpha_{31} \times B_1 \times t$ (19)

The ME coupling coefficient depends only on the thickness ratio η and the constitutive parameters. However, the output voltage depends also on the total thickness t of the device, the applied external magnetic field and the frequency. However, in the low frequency approximation ($\tan(ka)/ka \cong 1$ for $ka \rightarrow 0$) (i.e. frequency much lower than the electromechanical resonance frequency of the device), the output voltage is frequency independent. Experimental data indicate that the output voltage increases by a factor of 10 to 1000 if the device operates at the electromechanical resonance frequency [38].

Using relation $I = (V_{oc} - V) / Z$ as expected for a voltage source with internal impedance Z, we derived the impedance of the device as:

$$Z = \left(\frac{\eta t}{i\omega w l}\right) \frac{s_{11}^e (s_{11}^m \eta + s_{11}^e (1-\eta))}{\varepsilon_{33} s_{11}^e \left[\eta s_{11}^m + (1-\eta)s_{11}^e\right] - \left(d_{31}^e\right)^2 \left[(1-\eta)s_{11}^e + \eta s_{11}^m \left(1 - \frac{\tan ka}{ka}\right)\right]}$$

(20)

The internal impedance (Z) of the multiferroic source is mostly imaginary $Im(Z)$ with only a small real impedance (resistance) component $Re(Z)$ due to the imaginary parts of the permittivity and compliances. These imaginary parts correspond to electrical and mechanical losses and are of order 10^{-3} of the real parts [please see ref. 44]. We have estimated from relation (20) that the imaginary component of the impedance as $Im(Z) >> Re(Z)$ - the real part of the impedance, which is important for the operation of the sensor as the $Re(Z)$ causes Johnson noise. Moreover, from (20) we see that the impedance can be further minimised by increasing the operation frequency, increasing the length L of the sensor and by minimising the sensor thickness t (note that current in the device flows along the thickness, not the length of the bar, see figure 1).

157

MULTIFERROIC SENSOR FOR MAGNETIC RECORDING READ HEADS

One of the most common way computers and other electronic devices store data is by recording it onto a magnetic hard disk drive (HDD) – a multi $ billion industry. The functionality of modern magnetic recording read heads has been based for more than half a century on the magneto-resistive effects [45-49] such as: anisotropic magneto-resistive (AMR), giant magneto-resistance (GMR) and tunnelling magneto-resistance (TMR). These operate by changing the internal resistance of a read sensor stack when interacting with the stray field from the recorded bits of a magnetic recording medium. A dc current passes through the sensor stack either current in-plane (CIP) or perpendicular to the plane (CPP) and the change in the stack resistance is translated into a read signal as a voltage amplitude change $\Delta V = I \times \Delta R$. With the increase in the recording densities, the read head sensor stack must get thinner and more sensitive to smaller and smaller bit sizes. However, reducing the size of the sensor brings a number of complications: i) horizontal magnetic biasing is required to keep the sensing layer into a single domain state and the biasing becomes ineffective as the thickness of the permanent magnets is reduced; ii) the operation frequency of the read head is limited by the ferromagnetic resonance frequency of the sensing layer; iii) maintaining the signal to noise ratio large; iv) the construction of the sensor itself becomes very difficult for a sensor stack containing minimum 15 layers with a total thickness < 20 nm; vi) power consumption for portable devices and thermal stability of the reader; etc.

A possible solution is a new reader technology, which is based on the magneto-electric (ME) effect in MF materials rather than on the traditional magneto-resistance effects. This technology has been first proposed in 2007 [50,51] and experimentally demonstrated in 2008 [52]. In this paper we briefly examine the design requirements and the theoretical performance of such a MF reader when assuming a $1Tb/in^2$ recording medium. Our proposed design for a MF magnetic recording read head is shown in figure 2. The sensor stack is a planar structure that incorporates the tri-layer MF discussed in the previous section, with the addition of some extra layers. We assume that the total thickness and the width of the stack are again small in comparison to its length, so that the theoretical derivation for the tri-layer MF is valid for the full sensor stack. The operation of this MF read head is very simple. As the head moves along the magnetic recording track, the field from a recorded bit provides the ac excitation field to the MF read head sensor. The stress is induced via a dc magnetic bias field, which is provided by the interface exchange coupling of the FM layers with the anti-FM layers [53]. The use of exchange bias effect to create self-biased MF thin film structures is a very novel method for designing MF sensors. However, due to constrictions related to the exchange bias effect, such structures can only be created in thin film MF with FM thickness of maximum 20 – 30 nm, beyond which the exchange bias field becomes ineffective. Due to the linear relationship between the induced ac voltage and the applied field (see relation 19), the sensor's ac response voltage oscillates following the pattern of the recorded bits. The frequency of the read back signal is given by the relationship between the rotation speed of the disk and the linear recording density. In order to ensure the biasing of both FM electrodes, the proposed sensor stack consists therefore of 7 layers arranged as: Seed / AFM / FM / FE / FM / AFM / cap (see figure 2).

Theoretical simulations of the voltage output of the tri-layer MF introduced in the previous section (relation (18)) revealed that the optimum thickness ratio between the FM and FE is $t_e/t_m = 2$ (i.e. $\eta = 0.5$) assuming constant material parameters and field conditions throughout [50]. In order to calculate the output voltage of this reader (relation (18)) we now refer to the specs of a read head for data densities of 1 Tb/in^2, as listed in reference [54]: reader width 29 nm, bit length 11 nm, shield-to-shield spacing 22 nm, half gap thickness 5 nm, sensor stack thickness 12 nm.

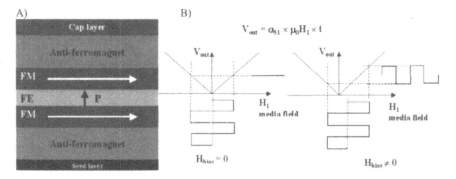

Figure 2. A) Multilayer stack design of a magnetic recording read head sensor based on the magneto-electric effect in multiferroics. The sensor generates a voltage output when subjected to the media ac magnetic field of the magnetic memory bits; B) Reader must be biased in order to sense the bit field (i.e. logical 0s and 1s). The biasing in this design is achieved using the exchange bias effect.

We also chose the sensor length (e.g. stripe height - as known in the magnetic recording industry) larger than the width and thickness so that it operates in the length expansion bar, L = 2a = 120 nm. In order to meet these requirements and based on previous modelling results [50], the individual thickness of each layer is: t_e = 3 nm, t_m = 1.5 nm, t_{seed} = 1 nm, t_{AFM} = 2 nm and t_{cap} = 1 nm, where the subscript refers to the specific layer. Assuming that the FE layer is PZT and that the FM layers are good conductive / magnetostrictive, the following parameters have been used for the calculations: s_{11}^e = 17×10^{-12} × (1-0.01j) m^2/N, d_{31}^e = -125×10^{-12} m/V, s_{11}^m = 2×10^{-11} × (1-0.01j) m^2/N, g_{11}^m = 3×10^{-3} A m/N, ε_{33} = 1750×ε_0, the tri-layer sensor thickness t = 6 nm and media field 100 Oe. In the low frequency approximation (tan(ka)/ka $\cong$ 1 for ka → 0), the reader output voltage is estimated as 5.7 µV. This can be greatly increased if the sensor operates at the resonance frequency, which is approximately given by ka = π/2. Assuming a 1% electromechanical loss in the computation, the calculated output voltage at resonance is 1.57 mV, which corresponds to a 3.14 mV peak-to-peak signal. The calculations indicate that when the sensor operates at the resonance frequency, the voltage response is insensitive to the sensor length, while the resonance frequency changes strongly with it. This result is very important as it suggests that the resonance frequency (i.e. data transfer rate) can be tuned via the sensor length. Unlike the existing magneto-resistive readers that are limited by the ferromagnetic resonance frequency of the permalloy thin films (~ 3 GHz), MF readers can operate at frequencies well beyond 10 GHz. The advantages of such a MF reader technology against the existing magneto-resistive readers are:

i)   The data is read back directly as an induced voltage and the dc test current used in magneto-resistive readers is no longer required. This reduces the power consumption by a small amount, which is important for portable device. Effectively the MF reader is self powered

ii)   The high resistance reader stack and the thermal performance are no longer a problem as the MF reader is capacitive rather than resistive

iii)   The frequency of operation is much higher resulting in faster read back speeds

iv)   The reader contains maximum 7 layers reducing the total thickness of the sensor and allowing higher recording densities to be achieved

v)   Horizontal magnetic biasing is no longer required as the domain state of the FM layers is no longer critical

vi)   A less complex structure containing only 7 layers and no horizontal biasing reduces the cost of production

CONCLUSIONS

We have discussed the advantages of using composite MF structures against single phase MF. Ideally, the most beneficial would be a single phase MF material that shows extremely large ME coupling effects and operates at room temperature. Such an important discovery would make a huge impact but so far the industrial requirements for applications have been fulfilled by MF composites.

The issue of ME coupling in the case of direct or strain mediated coupling has been also discussed. We showed that regardless of the type of coupling, the ME coupling coefficient is derived identically, with important implications for device functionality and measurement techniques.

A simple tri-layer MF laminated structure has been discussed and the analytical relation for the magnetically induced voltage of such a structure has been detailed. We also tabulated the constitutive equations for all possible cases of independent variable of electro-mechanical MF structures, depending on the sensor geometry and external conditions.

Finally a magnetic recording read head technology based on MF composites has been introduced. The derived analytical were used to calculate the performance of such a reader when used in conjunction with a 1 Tb/in^2 recording medium. We showed that the proposed MF sensor brings considerable advantages and could be used in the future to replace the existing magneto-resistive technology.

ACKNOWLEDGMENTS

The authors would like to acknowledge the financial support to this project received from the UK Department of Trade and Industry under the "Multiphysical Probing of Multifunctional Materials" - NMS grant.

160

REFERENCES

[1] S. X. Dong, J. Y. Zhai, F. Bai, J. F. Li, and D. Viehland, Appl. Phys. Lett. 87, 062502 (2005)

[2] S. X. Dong, J. Zhai, J.-F. Li, and D. Viehland, Appl. Phys. Lett. 88, 082907 (2006)

[3] S. X. Dong, J. G. Bai, J. Y. Zhai, J. F. Li, G. Q. Lu, D. Viehland, S. J. Zhang, and T. R. Shrout, Appl. Phys. Lett. 86, 182506 (2005)

[4] S. X. Dong, J. F. Li, and D. Viehland, Appl. Phys. Lett. 84, 4188 (2004)

[5] S. Shastry, G. Srinivasan, M. I. Bichurin, V. M. Petrov, A. S. Tatarenko, Phys. Rev. B 70, 064416 (2004)

[6] Y. K. Fetisov and G. Srinivasan, Appl. Phys. Lett. 88, 143503 (2006)

[7] Z. Huang, J. Appl. Phys. 100, 114104 (2006)

[8] M. Gajek, M. Bibes, S. Fusil, K. Bouzehouane, J. Fontcuberta, A. Barthelemy, A. Fert, Nature Materials, 6 (2007) 296-302

[9] P. Curie 1894 J. Physique 3 393

[10] D.N. Astrov, J. Exxp. Theoret. Phys. (U.S.S.R) 38, 948 (1960)

[11] G.T. Rado, V.J. Folen, Phys. Rev. Lett. 7 (1961) 310

[12] V.J. Folen, G.T. Rado, E.W. Stader, Phys. Rev. Lett. 6 (1961) 607

[13] R. Ramesh, N.A. Spaldin, Nature Materials 6 (2007) 21

[14] U. Laletsin, N. Padubnaya, G. Srinivasan, C.P. Devreugd, Appl. Phys. A 78 (2004) 33

[15] T.R. McGuire, E.J. Scott, F.H. Grannis, Phys. Rev. 102 (1956) 1000

[16] J.P. Rivera, Ferroelectrics, vol 161 (1994) 165

[17] G.T. Rado, Phys. Rev. Lett. 13 (1964) 335

[18] J.P. Rivera, Ferroelectrics, vol 161 (1993) 147

[19] G.T. Rado, J.M. Ferrari, W.G. Maisch, Phys. Rev. B 29 4041

[20] B.B. Krichevtsov, V.V. Pavlov, R.V. Pisarev, JETP Lett. 49 (1989) 535

[21] J. Wang, J.B. Neaton, H. Zheng, V. Nagarajan, S.B. Ogale, B. Liu, D. Viehland, V. Vaithyanathan, D.G. Schlom, U.V. Waghmare, N.A. Spaldin, K.M. Rabe, M. Wuttig, R. Ramesh, Science vol 299 (2003) 1719

[22] J. Hemberger, Nature 434 (2005) 364

[23] M.E. Lines, A.M. Glass, Principles and Applications of Ferroelectrics and Related Materials, 1977

[24] H. Sugie, N. Iwata, K. Kohn, J. Phys. Soc. Japan, 71 (2002) 1558

[25] T. Kimura, T. Goto, H. Shintani, K. Ishizaka, T. Arima, Y. Tokura, Nature 426 (2003) 55

[26] N. Hur, S. Park, P.A. Sharma, J.S. Ahn, S. Guha, S-W. Cheong, Nature 429 (2004) 392

[27] D. Fox, J.F. Scott, J. Pfys. C 10 (1977) L329

[28] T. Kimura, G. Lawes, G.A.P Ramirez, Phys. Rev. Lett. 94 (2005) 137201

[29] V.R. Palkar, S.K. Malik, Solid State Com. 786 134 (2005) 783–786

[30] C.W. Nan, L.Liu, N. Cai, J. Zhai, Y. Ye, Y.H. Lin, L.J. Dong, C.X. Xiong, Appl. Phys. Lett. 81 (2002) 3831

[31] N. Cai, C.W. Nan, J. Zhai, Y. Lin, Appl. Phys. Lett. 84 (2004) 3516

[32] J Ryu, V Carazo, K. Uchino, H.E. Kim, Jpn. J. Appl. Phys. 40 (2001) 4948

[33] G. Srinivasan, E. T. Rasmussen, B. J. Levin, and R. Hayes, Phys. Rev. B 65 (2002) 134402

[34] M.K. Lee, T.K. Nath, C.B. Eom, M.C. Smoak, F. Tsui, Appl. Phys. Lett. 77 (2000) 3547

[35]   S.S. Kim, J.W. Lee, S.C. Shin, H.W. Song, C.H. Lee, K. No, J. Magn. Magn. Mater. 267 (2003) 127
[36]   G. Srinivasan, E. T. Rasmussen, J. Gallegos, R. Srinivasan, Yu. I. Bokhan, and V. M. Laletin, Phys. Rev. B 64 (2001) 214408
[37]   H. Ryu, P. Murugavel, J.H. Lee, S.C. Chae, T.W. Noh, Y.S. Oh, H.J. Kim, K.H. Kim, J.H. Jang, M. Kim, C. Bae, J.G. Park, Appl. Phys. Lett. 89 (2006) 102907
[38]   U. Laletsin, N. Padubnaya, G. Srinivasan, C.P. Devreugd, Appl. Phys. A 78 (2004) 33
[39]   G. Srinivasan, E. T. Rasmussen, R. Hayes, Phys. Rev. B 67 (2003) 014418
[40]   M. Fiebig, , Revival of the magnetoelectric effect, J. Phys. D: Appl. Phys. 38 (2005) R123
[41]   W. Eerenstein, N.D. Mathur, J.F. Scott, Nature 442 (2006) 759
[42]   M. Vopsaroiu, M. Stewart, T. Hegarty, A. Muniz-Piniella, N. McCartney, M. Cain, G. Srinivasan, Meas. Sci. Technol. 19 (2008) 045106
[43]   J. Blackburn, M. Vopsaroiu, M. G. Cain, Journal of Applied Physics, 104 074104 (2008)
[44]   J.F. Blackburn, M.G Cain, J. Appl. Phys. 100 (2006) 114101
[45]   S. Tumanski, Thin Film Magnetoresistive Sensors, IOP publishing (2001) ISBN 0750307021
[46]   Sarah M Thompson, The discovery, development and future of GMR: The Nobel Prize 2007, J. Phys. D: Appl. Phys. 41 (2008) 093001
[47]   E. M. Williams, Design and Analysis of Magnetoresistive Recording Heads, (2001) ISBN 0-471-36358-8
[48]   M. Julliere Phys. Lett. 54A (1975) 225
[49]   J. C. Slonczewski, Phys. Rev. B 39 (1989) 6995
[50]   M. Vopsaroiu, J. Blackburn, M. G. Cain, J. Phys D: Applied Physics, Vol. 40, pp. 5027 (2007)
[51]   M. Vopsaroiu, J. Blackburn, A. Piniella, M. G. Cain, Journal of Applied Physics 103, 07F506 (2008)
[52]   Y. Zhang, L. Zheng, C. Deng, J. Ma, Y. Lin, Nan Ce-Wen, Applied Phys. Letters 92, 152510 (2008)
[53]   E. Fulcomer, S. H. Charap, J. Appl. Phys. 43, 4184 (1972)
[54]   R. Wood, J. Miles, T. Olson, IEEE Trans. Magn. Vol. 38, No 4, 1711 (2002)
[55]   H. Zheng, J. Wang, S.E. Lofland, Z. Ma, L. Mohaddes-Ardabili, T. Zhao, L. Salamanca-Riba, S.R. Shinde, S.B. Ogale, F. Bai, D. Viehland, Y. Jia, D.G. Schlom, M. Wuttig, A. Roytburd, R. Ramesh, Science 303 (2004) 661-663
[56]   R.N.P. Choudhury, C.odríguez, P. Bhattacharya, R.S. Katiyar. C. Rinaldi, Journal of Magnetism and Magnetic Materials, Volume 313, Issue 2 (2007) 253-260
[57]   J. Zhai, N. Cai, Z. Shi, Y. Lin, Ce-Wen Nan, J.Phys. D: Appl. Phys. 37 (2004) 823-827
[58]   Shuxiang Dong, Jie-Fang Li, D. Viehland, Appl. Phys. Lett. 83 (2003) 2265-2267

Mater. Res. Soc. Symp. Proc. Vol. 1161 © 2009 Materials Research Society   1161-I07-04

Room Temperature Ferromagnetism and Lack of Ferroelectricity in Thin Films of 'Biferroic?' YbCrO$_3$

Sandeep Nagar[1], K. V. Rao[1], Lyubov Belova[1], G. Catalan[4], J. Hong[4], J. F Scott[4], A. K Tyagi[3], O. D Jayakumar[3], R. Shukla[3], Yi-Sheng Liu[2] and Jinghua Guo[2]

[1]Dept. of Materials Science, Royal Institute of Technology, Stockholm, Stockholm Ian, Sweden

[2]Advanced Light Source, Lawrence Berkeley National Laboratory, Berkeley, California

[3]Chemistry Division, Bhabha Atomic Research Centre, Mumbai, India

[4]Department of Earth Sciences, University of Cambridge, Cambridge, United Kingdom

ABSTRACT

Search for novel multi-functional materials, especially multiferroics, which are ferromagnetic above room temperature and at the same time exhibit a ferroelectric behavior much above room temperature, is an active topic of extensive studies today. Ability to address an entity with an external field, laser beam, and also electric potential is a welcome challenge to develop multifunctional devices enabled by nanoscience. While most of the studies to date have been on various forms of Bi- and Ba based Ferrites, rare earth chromites are a new class of materials which appear to show some promise. However in the powder and bulk form these materials are at best canted antiferromagnets with the magnetic transition temperatures much below room temperature. In this presentation we show that thin films of YbCrO$_3$ deposited by Pulsed Laser Deposition exhibit robust ferromagnetic properties above room temperature. It is indeed a welcome surprise and a challenge to understand the evolution of above room temperature ferromagnetism in such a thin film. The thin films are amorphous in contrast to the powder and bulk forms which are crystalline. The magnetic properties are those of a soft magnet with low coercivity. We present extensive investigations of the magnetic and ferroelectric properties, and spectroscopic studies using XAS techniques to understand the electronic states of the constituent atoms in this novel Chromite. While the amorphous films are ferromagnetic much above room temperature, we show that any observation of ferroelectric property in these films is an artifact of a leaky highly resistive material.

INTRODUCTION

Perovskite oxides show an interesting functionality known as magnetoelectricity, in which coupling between electric polarization and magnetism could in principle lead to new applications in spintronics and memory storage [1, 2]. On the other hand, magnetoelectric multiferroics (materials which are simultaneously magnetic and ferroelectric above room temperature, and with coupling between these two properties) are extremely rare [3]. A potentially promising family of materials are the rare-earth orthochromites. These are antiferromagnetic insulators with a weak canting moment due to the Dzhialoshinskii-Moriya interaction [5, 6, 7]. Moreover, their magnetic symmetry is such that it allows for linear magnetoelectric coupling (electrically-induced magnetization or magnetically induced polarization) [7]. They are not, however, multiferroic, because they are centrosymmetric and therefore cannot be ferroelectric.

In spite of their centrosymmetry, it has been claimed that rare-earth orthochromites may actually be ferroelectric [8, 9, 10]. The experimental evidence presented so far, however, is compatible with conductive artifacts since orthochromites are not very good insulators [11]. On the other hand, materials that are centrosymmetric in bulk may indeed become ferroelectric in thin film form due to epitaxial strain [4], so there is a chance that some orthochromites may become ferroelectric under strain. In order to explore these issues we have investigated here the electric and magnetic properties of $YbCrO_3$. When grown as amorphous thin films, we have observed not only an enhanced ferromagnetism, but also shows excellent ferroelectric-like hysteresis which, however, in reality is due to leakage currents.

EXPERIMENT

The $YbCrO_3$ powder was prepared by a combustion reaction between Ytterbium nitrate, chromium nitrate using glycine as a fuel. The reactants were mixed in the required molar ratios which on thermal dehydration resulted in formation of viscous liquid (gel). On increasing the temperature (» 250 °C) the viscous liquid swelled and auto-ignited. To remove traces of carbon impurities (if any), the powders were calcined at 600°. The fuel-deficient ratio (1:0.50) resulted in the formation of $YbCrO_4$ after a flameless combustion, which was further heated at 800°C for 1 h to get $YbCrO_3$. Sufficient powder was then Surface Plasma Sintered to make a pellet with density of 99.5% which was used as a target for Pulsed LASER deposition technique. The thin films were deposited by means of Pulsed LASER deposition using a Continuum Laser in the 355nm mode at 10 Hz and energies of the order of 200 mJ per pulse. Thickness of films were determined by sectioning the films using a dual beam Focused Ion Beam (FIB) facility NOVA 600. Films were deposited in oxygen and nitrogen atmospheres on silicon, and sapphire substrates kept at 300°C.

DISCUSSION

Structural characterization:

Thin films were characterized by XRD and TEM for their crystal structure, both of which confirm amorphous structure (figure 1a,1c). X-Ray Diffraction graph does not show any peak for $YbCrO_3$ crystal formation in any of our films. Also TEM halo indicates amorphous thin film. Energy Dispersive Spectroscopy was performed along with TEM investigation which confirmed that we do not have any kind of impurity even at atomic level in these samples. Also using Focused Ion-Beam assisted Scanning Electron Microscope (figure 1c); we confirmed a continuous film of thickness ≈ 243nm.

Figure 1: *(a) TEM diffraction pattern of thin film shows a halo for diffraction pattern indicating complete amorphization of the thin film. (b) Using Focused Ion-Beam assisted Scanning Electron Microscope; thickness of film was calculated to be around 243nm. (c)XRD of YbCrO3 pellet (used as target) orthorhombic crystal structure and thin films show only peaks for silicon substrate indicating amorphous nature.*

Magnetic characterization:

The ceramic pellet shows no remnant magnetization (figure 2) at any temperature, consistent with the antiferromagnetic symmetry expected for crystalline YbCrO3. The thin films (figure 2), in contrast, show hysteresis with a saturation magnetization of 1.26 emu/g. Since our films are deliberately amorphous, strain should not affect their functional properties –including magnetism- in any way. Also, the lack of symmetry means that the behaviour should be dominated by short range interactions. The Cr^{+3}-O^{2-}-Cr^{+3} super-exchange interaction is antiferromagnetic in nature, so the observation of remnant magnetization in this context is surprising, and suggests that the origin must be linked to ions other than Cr^{3+}.

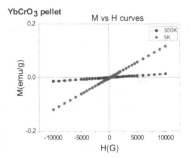
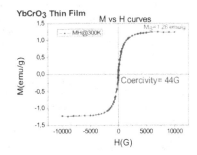

Figure 2: *Comparison between magnetic hysteresis for crystalline ceramics (left) and amorphous films (right).*

In order to elucidate this, we have performed XAS measurements (Figure 3) The XAS results reveal a mixture of Cr^{+3} and Cr^{+4} ions. Yb is one of the few lanthanides which is able to become divalent. Since the valence state of the Yb depends on that of its surrounding ions, it is possible that in the amorphous structure of our thin films, some Yb ions are more stable in the Yb^{+2} valence state, which of course must be compensated by Cr^{4+} in order to preserve charge neutrality.

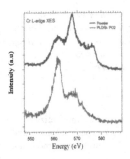

Figure 3: *The XAS spectra of the thin films indicates a mixture of Cr^{+3} and Cr^{+4}*

In stoichiometric crystalline $YbCrO_3$, the magnetic symmetry is mostly dominated by antiferromagnetic superexchange, as mentioned earlier. This means that the localized d-shell electrons of the Cr^{+3} ions align parallel to the $2p$ electrons of the oxygen O^{2-}; since the two electrons in the p orbital of the oxygen are anti-parallel, the spins of the Cr^{+3} ions will be anti-parallel to each other, giving net antiferromagnetic order. This, obviously, requires that the electrons be localized, and indeed antiferromagnetic oxides are always insulators, as is the case for $YbCrO_3$.

But, in case of mixed valence states of the metal ions, ferromagnetic double-exchange interaction dominates, whereby an electron hops between the two metal ions. Since the spin of the electron does not change during the hopping, this interaction tends to align the metal ions ferromagnetically [15]. Thus, mixed valence, ferromagnetism and electron delocalization (increased conductivity) tend to go together in magnetic oxides. This, we believe, may also be the case for our chromite thin films, so that the electric, magnetic and chemical measurements are consistent with each-other: the appearance of ferromagnetism is correlated with the mixed valence of the Cr ions, while the "ferroelectric" hysteresis loops are due to enhanced conductivity, itself consistent with ferromagnetic double-exchange.

Electrical characterization:

A room temperature P(E) loop measured at 1 kHz (figure 4a) is indistinguishable from a genuine ferroelectric hysteresis loop; due to conduction artefacts. If the sample is not a perfect insulator, then the application of voltage induces also leakage currents which are also integrated by the ferroelectric tester [13]. The integrated charge can look very similar to true ferroelectric hysteresis [12] and leakage current will be also be integrated for measurements [13].

We employed two methods to investigate this phenomenon further. Firstly, since hysteresis loops are essentially time independent so any change in shape of loop will indicate leakage currents. This is indeed the case with our samples, as shown in figure 4b.

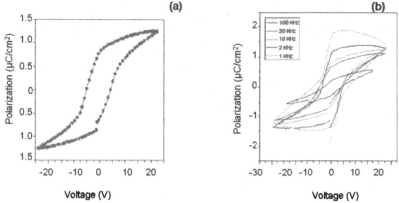

Figure 4: *(a) P(E) measurement at 1KHz shows "ferroelectric like" loop (b)Hysteresis loops at different frequencies*

The second method is to perform pulsed measurements (called PUND, which stands for "positive-up, negative-down) which allow separating the switchable from the non-switchable part of the polarization. The first pulse of each pair is the "switching" pulse, and it integrates both switching and non-switching charge, while the second pulse only integrates non-switching charge (figure 5). Subtracting both, then, the true remanent polarization can be revealed. PUND measurement in our case reveals zero remnant magnetization which shows the "apparent polarization" is due to charge leakage only.

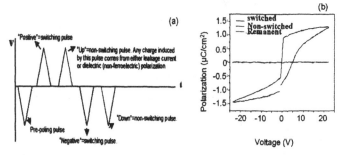

Figure 5: *(a) Schematic of a PUND measurement. (b) PUND measurement of a YbCrO₃ amorphous film.*

CONCLUSIONS

In summary, the present work illustrates several interesting effects. First, that "pseudo-hysteresis" loops can be obtained in a non-ferroelectric, amorphous semiconductor. Second, that

167

the magnetic and transport properties of the orthochromite depends on the valence state of the Cr ions; mixed valence in the Cr ion leads to ferromagnetism and increased conductivity, both of which are indicative of local double-exchange interactions. And third, that the mixed valence of the Cr ions is itself consistent with the fact that Yb ions are allowed to be in the Yb^{2+} state. We also note here that, although the mixed valence state of the transition metal in our films is itself likely to be due to the mixed valence nature of the Yb in a glassy environment, this is by no means an essential requirement: oxygen vacancies in a perfectly ordered pervoskite structure would also lead to mixed valence in order to preserve charge neutrality. While here we have used $YbCrO_3$ as a case study, in fact the mechanisms discussed for the interplay between magnetic, transport and ferroelectric-like properties are general and should apply to other transition metal oxides such as manganites or ferrites.

ACKNOWLEDGMENTS

It is pleasure to thank Prof. Mats Nygren for using their SPS technique at the Stockholm University. Research in Sweden is supported by the Funding Agency VINNOVA, Swedish Science Research Council, and the HERO-M Centre of Excellence at KTH. The Advance Light Source at Berkeley is supported by the Director, Office of Science, Office of Basic Energy Sciences, of the U.S. Department of Energy under Contract No. DE-AC02-05CH11231. Work done at BARC, India is supported by the Atomic Energy Commission, India.

REFERENCES

1.   M. Fiebig, J. Phys. D, **38**, R123(2005).

2.   W. Eerenstein, N. D. Mathur, and J. F. Scott, Nature, **442**, 759 (2006).

3.   N. A. Hill, J. Phys. Chem., **B104**, 6694 (2000).

4.   J. H. Haeni et al, Nature, **430**, 758-761 (2004)

5.   Shtrikman, S., Wanklyn, B. M., Yaeger, I., Int. J. Magn. 1 327 (1971).

6.   N. Kojima and I. Tsujikawa, J. De Physique **49**, C8-897 (1988).

7.   T. Yamaguchi and K. Tsushima, Phys. Rev. B **8**, 5187 (1973).

8.   G.V. Subba Rao, G.V. Chandrashekhar and C.N.R. Rao, Solid State Communications, **6** 177 (1968).

9.   C. R. Serrao, A. K. Kundu, S. B. Krupanidhi, U. V. Waghmare and C. N. R. Rao, Phys. Rev. B **72**, 220101R (2005).

10. J. R. Sahu, C. R. Serrao, N. Ray, U. V. Waghmare and C. N. R. Rao, J. Mat. Chem. **17**, 42 (2007)

11. G. V. Subba Rao, B. M. Wanklyn And C. N. R. Rao, J. Phys. Chem. Sol. **32**. 345 (1971).

12. L. Pintilie, M. Alexe, App. Phys. Lett. 87 (11): 112903 (2005).

13. J. F. Scott, J. Phys: Cond. Matt. **20**, 21001 (2008).

14. G. Catalan, J. F. Scott, Nature **448** (7156): E4-E5 (2007).

15. C. Zener, Phys. Rev. **82**, 403 (1951)

Mater. Res. Soc. Symp. Proc. Vol. 1161 © 2009 Materials Research Society 1161-I07-05

Competing Magnetic Interactions in Magnetoelectric YbMnO$_3$

Shishir K. Ray, Ying Zou, Mark S. Williamsen, Somaditya Sen, Larry Buroker,
and Prasenjit Guptasarma

Department of Physics, University of Wisconsin - Milwaukee, Milwaukee, WI 53211

ABSTRACT

We present studies of magnetization and heat capacity of a single crystal of YbMnO$_3$ in variable temperature and magnetic field, and clarify several new aspects of the magnetic field-temperature phase diagram. YbMnO$_3$ is a rare-earth manganite oxide with hexagonal crystal symmetry in which two multiferroic ordered states – ferroelectricity and antiferromagnetism – coexist at low temperature. Single crystals of YbMnO$_3$ were carefully grown from a Floating Zone (FZ) at low speed, then oriented and studied with the magnetic field oriented along the c-axis. Magnetization and heat capacity measurement show features corresponding to long range anti-ferromagnetic (AFM) ordering of Mn^{3+}, and the rare earth Yb^{3+}. The ordering temperature of Mn^{3+} is independent of applied magnetic field up to 5T. However, contrary to previous reports in flux-grown crystals, we do not observe a complete suppression of Yb^{3+} order above 0.1T. Instead, we find that Yb^{3+} remains ordered at least up to 1 T, suggesting a revision of our current understanding of the ordering mechanism of the Mn-Yb and Yb –Yb sub-lattices in this hexagonal structure.

INTRODUCTION

Magnetoelectric rare earth manganite oxides (RMnO$_3$) are a rare, relatively new, and fascinating class of materials in which ferroelectricity and (anti)ferromagnetism coexist within the same crystal phase [1,2]. The two ordered states are found to be coupled: applied magnetic field can change ferroelectric polarization, and an applied electric field can change magnetization, making them suitable for applications such as switching.[3] The hexagonal manganite YbMnO$_3$ has received considerable attention in recent years due to the presence of ferroelectricity at room temperature (T$_c$ = 973 K), low temperature magnetic properties and possible frustrated magnetic order (T$_N^{Mn}$ = 80 K, T$_N^{Yb}$= 4K). The fascinating magnetoelectric properties at low temperature in hexagonal RMnO$_3$ is believed be due to the rare earth atom being shifted from its central position and surrounded by triangular AFM order [4-9].

EXPERIMENT

A single crystal of YbMnO$_3$, grown by us from a floating zone, was used for all measurements reported here. Starting materials for the feed and seed rods for single crystal growth were fabricated by conventional solid state reaction of high-purity Yb$_2$O$_3$ and MnO$_2$. Precursor oxides were mixed in stoichiometric proportions, ground in an agate ball mill, and reacted at a final temperature of 1200°C for 24 hrs in O$_2$ after intermediate iterations of grinding and heating. The resulting single phase powder was packed inside silicone molds and pressed at 45 kPsi in an isostatic press, yielding high-density pressed rods 70mm in length and 6mm in diameter. The rods were subsequently sintered in Argon at 1200°C for 24hrs. Powder x-ray diffraction (PXRD) data, obtained from a SCINTAG 2000 diffractometer in the θ-2θ mode, is shown in Fig. 1. Rietveld refinement, performed using TOPAS Academic V4.1, reveals a crystal structure with hexagonal symmetry, and lattice parameter values of a=6.0700(2) and c= 11.397(4), in agreement with previously reported values for YbMnO₃ [10].

The single crystal was grown from a floating zone (FZ) in an NEC SC I-MDH-11020 two-mirror image furnace, with a growth rate of 1.1 mm/hr in a gas mixture of (0.2)O₂-(0.8) flowing at 1 atm.[11] The crystal was cut, polished, and oriented using back-reflection geometry Laue diffractograms. Selected area electron diffraction (SAED) and High Resolution Transmission Electron Microscopy (HRTEM) confirm the symmetry and homogeneity of the crystal structure, shown as an inset in Fig.1(b). Chemical stoichiometry and crystal surface morphology were studied using Scanning Electron Microscopy and Energy Dispersive x-ray (EDX) analysis. Powder x-ray diffraction on crushed single crystals revealed a small peak attributable to Yb₂O₃.

Magnetization in ac- and dc- modes, and heat capacity measurements were carried out in a Physical Property Measurement System (PPMS) and a Magnetic Property Measurement System (MPMS) by Quantum Design. In all cases, the magnetic field was applied parallel to the c-axis.

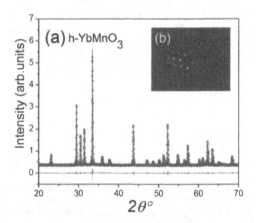

Figure 1. Crystal Structure & Symmetry: (a) Rietveld analysis of the powder x-ray diffraction pattern of YbMnO₃, as described in the text reveals P63cm symmetry. The fit is shown as a line connecting the data points; with the residue shown at the bottom. (b) Selected area electron diffraction (SAED) pattern obtained from a single crystal of YbMnO₃, showing hexagonal symmetry.

RESULTS & DISCUSSION

DC Magnetization

DC magnetic susceptibility (χ) of YbMnO₃ was measured using a Magnetic Property Measurement System (MPMS) from Quantum Design, in the temperature range of 2-300K, with the externally applied dc field aligned along the c-axis of the single crystal. Fig.2 shows inverse susceptibility $1/\chi$ as a function of temperature. The extrapolated high temperature linear fit indicates a negative intercept around -180 K, which is close to the value of -175K reported by others in the same system with applied field along the c-axis, and indicates anti-ferromagnetic (AFM) order. [11-15]. The ratio of $|\theta|$ and T_N in this system is a low ~ 3. This is especially interesting when compared with reported values of ~8-10 for other hexagonal manganite systems such as $ScMnO_3$, $YMnO_3$ and $LuMnO_3$ [10]. A large $|\theta| / T_N$ ratio is generally believed to be an indicator of a spin system with geometrical frustration because it is difficult to explain such a high value from elementary mean field theory [16]. This indicates that, unlike other hexagonal manganites, $YbMnO_3$ does not display significant spin frustration.

Magnetic DC susceptibility was measured for both field-cooled (FC) and zero-field cooled (ZFC) conditions, with a field of 100 Oe applied along the c-axis as shown in Fig 2(b). It is interesting to note that, in our measurements, ZFC and FC follow each other very closely from high temperature down to ~4K where they separate. This is contrary to previous measurements in flux-grown crystals [10,13] in which ZFC and FC are found to separate below the Mn ordering transition temperature ~80K. This is believed to be an indication of an unconventional order between Mn moments, and a possible lack of compensation below T_N[13]. Fontcuberta et al [13] also find that the temperature at which ZFC and FC separate decreases with increasing applied field above 1kOe. In our single crystal, which was grown from a floating zone, ZFC and FC do not separate below ~80K, even in a very low field of 100Oe, pointing to the fact that Mn is well-ordered in floating-zone grown single crystals, generally considered to be of higher quality and homogeneity.

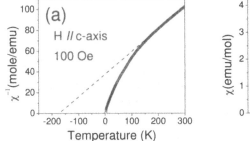 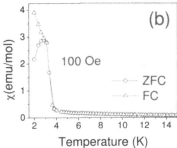

Figure 2. (a) Inverse DC Magnetic Susceptibility ($1/\chi$) of a floating-zone grown $YbMnO_3$ single crystal as a function of temperature, with an externally applied field of 100 Oe. (b) DC Magnetic susceptibility of $YbMnO_3$ measured as a function of temperature, with applied magnetic field along c-axis. The circles indicate susceptibility when the sample is zero field cooled (ZFC). The triangles represent a field cooled (FC) measurement with field of 100 Oe.

AC Magnetization

Complex ac-susceptibility was measured with an applied external dc field varying between 0.01 and 5 Tesla along the c-axis of the $YbMnO_3$ single crystal. The applied ac field was kept constant at 2 Oe, with a driving frequency of 1 kHz. Fig. 3 shows ac susceptibility measured with a dc field of 100 Oe, and with temperature varying between 2K and 45K. A feature is seen at ~3.5 K, indicating an ordering of the rare-earth ion Yb^{3+}. This is clearly visible in both in and out of the plane components of ac-susceptibility. Note that the out of plane component remains very stable and close zero for the entire temperature range, indicating a stable AFM ordered Mn lattice. AC susceptibility measured as a function of temperature at different external applied field (0.01 – 5 T) is shown in Fig. 4 below.

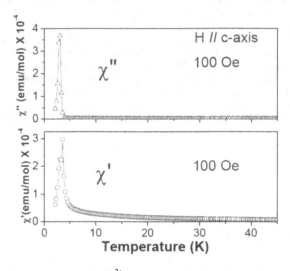

Figure 3. Real (χ') and imaginary (χ'') component of AC magnetic susceptibility plotted as a function of temperature, with magnetic field appied along the c-axis of the single crystal.The driving AC-frequency is 1 kHz and the AC- field is 2 Oe.

An ordering of the rare-earth Yb^{3+} ion at low temperature is clearly observed at all values of applied field. The T_N^{Yb3+} peak appears at ~3.5 K and shifts to higher temperature with increasing applied field. It is important to note here that, unlike previous reports, we clearly observe the rare-earth ordering peak up to an applied field of 1 T.The peak then shifts toward higher temperature, becoming sharper with increasing applied field, as observed in the 5T data.

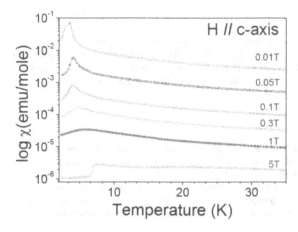

Figure 4. AC-susceptibility as a function of temperature at different DC magnetic fields applied along c-axis of the single crystal of $YbMnO_3$. The AFM ordering feature at low temperature continues to exist up to 5T. The data for fields of 100 Oe and 500 Oe data were multiplied with scale factors. The vertical axis is presented as log χ to accommodate the entire data set in a single plot.

Heat Capacity with varying applied magnetic field:

Heat capacity of the YbMnO₃ single crystal was measured using a relaxation calorimetric technique in the range of 2-100 K with varying applied magnetic field. Fig. 5 shows a λ- type anomaly at $T_N\sim80$K, with varying applied magnetic field. The Mn^{3+} ordering peak at 80K is clearly visible at all the fields. There is no significant change in T_N^{Mn3+} ordering temperature with applied magnetic field.

The low temperature rare-earth ordering peak is also observable. Consistent with our magnetization measurement with variable field, and contrary to previous reports [12, 15] from flux-grown single crystals, heat capacity measurements clearly show the presence of long range AFM order of Yb^{3+} ions at 1T. This is in strong contradiction with reported values [12, 15], indicating that rare-earth order does not quench with low applied field. Based upon these observations, and together with our magnetization measurements, we believe that the H-T (magnetic field-temperature) phase diagram for YbMnO₃ needs revision. Detailed measurements with other values of applied magnetic field, along with a complete phase diagram, will be reported in a subsequent publication.

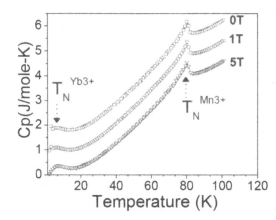

Figure 5. Specific heat study of YbMnO₃ as a function of temperature at three different magnetic fields between 0-5 T. The Yb^{3+} ordering, indicated with an arrow, is clearly visible at low temperatures at all fields, including at 1.0 T. The 0 T and 1.0 T data have been shifted slightly along the vertical axis for clarity of presentation.

CONCLUSIONS

We have grown high quality single crystals of YbMnO₃ from a floating zone, and studied the nature of long range anti-ferromagnetic order of Mn and Yb ions in this system using a combination of ac magnetization, dc magnetization and heat capacity measurements with varying magnetic field. We find that Yb^{3+} and Mn^{3+} moments order as observed in previous studies [17]. However, we find several differences with previously reported observations in flux-grown single crystals, which we attribute to increased homogeneity in our crystals, grown from floating zone. We find that the Yb^{3+} ordering peak is not suppressed upon increasing applied magnetic field up to 1 T, in contrast with previous reports [15] that this order is quenched above 0.07 T, suggesting that the H-T phase diagram of YbMnO₃ needs further revision. Based on our measurement of the ratio of $|\theta| / T_N$, we conclude that YbMnO$_3$ does not display significant spin frustration. This contradicts previous belief [12, 15] that members of the hexagonal family of RMnO$_3$ systems display geometric frustration in antiferromagnetic order. We suggest that the observation of spin frustration could result from crystalline disorder or inhomogeneity, and plan to extend our studies to other hexagonal manganite systems grown from a floating zone. Finally, our dc-magnetization measurements do not show a deviation between the zero-field cooled and field-cooled traces with temperature below Mn ordering, indicating that Mn spins are well-compensated in YbMnO$_3$, which we believe is a result of higher crystalline order in our single crystal.

ACKNOWLEDGMENTS

We thank the National Science Foundation, DMR-0449969, DMR-0509691 and DMR-0621445, a Research Growth Initiative award from the University of Wisconsin Milwaukee, and a Research Infrastructure Award from the Wisconsin Space Grant and NASA for support. We are grateful to D. Robertson, S. Hardcastle and A Skliarov for facilities support.

REFERENCES

*   Author for correspondence, email: pg@uwm.edu.
[1]   T. Kimura, T. Goto, H. Shintani, K. Ishizaka, T. Arima, and Y. Tokura, *Nature* 426, 55 (2003).
[2]   N. A. Hill, *J. Phys. Chem.* B 104, 6694 (2000).
[3]   S. Lee, *et al*, *Nature* 451, p.8052808 (2008).
[4]   M. Fiebig, T. Lottermoser, T. Lonkai, A. Goltsev, and R. Pisarev, *J. Magnetism and Magnetic Mater.* 290, 883 (2005).
[5]   M. Fiebig, D. Frohlich, K. Kohn, S. Leute, T. Lottermoser, V. V. Pavlov, and R. V. Pisarev, *Phys. Rev. Lett.* 84, 5620 (2000).
[6]   M. Fiebig, T. Lottermoser, and R. V. Pisarev, *J. of Appl. Phys.* 93, 8194 (2003).
[7]   U.Adem *et al* arXiv:0811.4547v1 [cond-mat.mtrl-sci]
[8]   B. Lorenz *Phys. Rev.* B 71, 014438 (2005)
[9]   I.Munawar et al *J. Phys.: Cond. Matt.* 18, 9575-9583 (2006)
[10]   T.Katsufuji, S.Mori,M.Masaki,Y.Moritomo,N.Yamamoto and H.Tagaki, *Phys. Rev.*B 64 104419 (2001)
[11]   S.K.Ray et al "Low speed growth of single crystals of Hexagonal Manganites with Float zone technique" (unpublished).
[12]   H. Sugie, N. Iwata, and K. Kohn, *J. Phys. Soc.Jap.* 71, 1558 (2002).
[13]   J. Fontcuberta, M. Gospodinov, and V. Skumryev, *J. Appl. Phys.* 103 (2008).
[14]   Y.H. Huang, H. Fjelivag, M. Karppinen, B.C. Hauback, H. Yamauchi, and J.B. Goodenough, *Chemistry of Materials* 18(8) pp. 2130-2134 (2006).
[15]   F. Yen, C. dela Cruz, B. Lorenz, E. Galstyan, Y. Y. Sun, M. Gospodinov, and C. W. Chu, *J. of Mater. Res.* 22, 2163 (2007).
[16]   P. Schiffer and A.P.Ramirez, *Comments in Cond. Matter Phys*, Phys. 18, 21 (1996).
[17]   X. Fabreges, I. Mirebeau, P. Bonville, S. Petit, G. Lebras-Jasmin, A. Forget, G. Andre, and S. Pailhes, *Phys. Rev.* B 78, 214422 (2008).

Printed in the United States
By Bookmasters